Crossing
THE LINES
We Draw

Faithful Responses to a
Polarized America

MATTHEW TENNANT

Foreword by Deidra Riggs

JUDSON PRESS
PUBLISHERS SINCE 1824
VALLEY FORGE, PA

Judson Press has made every effort to trace the ownership of all quotes. In the event of a question arising from the use of a quote, we regret any error made and will be pleased to make the necessary correction in future printings and editions of this book.

Unless otherwise indicated, Scriptures are the author's translation. Scriptures also from the New American Standard Bible, ©1960, 1962, 1963, 1968, 1971, 1972, 1973, 1975, 1977 by The Lockman Foundation. Used by permission. The HOLY BIBLE, New International Version®, NIV®, copyright ©1973, 1978, 1984, 2011 by Biblica Inc. Used by permission. All rights reserved worldwide. The Holy Bible, New Living Translation, copyright ©1996. Used by permission of Tyndale House Publishers, Inc., Wheaton, IL 60189. All rights reserved. The New Revised Standard Version of the Bible, copyright ©1989 by the Division of Christian Education of the National Council of the Churches of Christ in the United States of America. Used by permission. All rights reserved.

Interior and cover design by Wendy Ronga, Hampton Design Group.

Library of Congress Cataloging-in-Publication data

Names: Tennant, Matthew, 1973- author.
Title: Crossing the lines we draw: faithful responses to a polarized America/ Matthew Tennant; foreword by Deidra Riggs. Description: Valley Forge, PA: Judson Press, 2020. Identifiers: LCCN 2019046339 (print) | LCCN 2019046340 (ebook) | ISBN 9780817018122 (paperback) | ISBN 9780817082116 (epub) Subjects: LCSH: United States—Church history. | Christianity and culture— United States. Classification: LCC BR526 .T46 2020 (print) | LCC BR526 (ebook) | DDC 621/.10973—dc23
LC record available at https://lccn.loc.gov/2019046339
LC ebook record available at https://lccn.loc.gov/2019046340

Printed in the U.S.A.
First printing, 2020.

This book is dedicated to my sons, Dean and Edison.
You are the agents of reconciliation I hope to be.

CONTENTS

Foreword

Traveling in the back seat of our parents' car, on road trips from Michigan to Virginia, my little sister and I were always well-behaved. We sat upright and still, and we were quiet—both of us deep in thought. The entire twelve-hour trip passed like a mere moment in time as we patiently watched the miles fall away. From time to time, our parents would turn and smile at us from their perch in the front seat of the car.

Of course, none of this is true.

Those twelve-hour road trips tested the bounds of our parents' love for us, as (long before seatbelt laws) my sister and I conspired to build forts with blankets, squeeze our small bodies into the "rear deck" beneath the rear windshield for naps, beg truckers to blast the horns on their gigantic rigs, and kick the backs of our parents' seats. We sang loudly, we complained endlessly, and we begged for a bathroom break nearly every five miles.

By the *second* hour of our road trip, however, my sister and I had had enough of one another. One of us would sing the wrong words to a song or eat the last potato chip or skip a letter in the Alphabet Game. Our frustration with each other bubbled over into tears, and we retreated to opposite sides of the back seat.

"You stay over there!" one of us would yell to the other, pointing an indignant finger for emphasis. Then, just to make sure we were clear, we would divide the back seat in half with an imaginary line and forbid the other to dare to get close to it.

Sound familiar?

Whether you grew up taking road trips with siblings or not, I imagine you've found yourself at some point in time on one side of an imaginary line, while someone you know and (used to?) love is positioned on the other side of that line. Opportunities to "choose sides" seem to present themselves on a daily basis. Are you pro-life or pro-choice? Democrat or Republican? Conservative or Progressive? Do you read the Bible literally, or "liberally"? Is your church welcoming and affirming, or not? Did you vote for Trump or Hillary, or not at all? Do you eat gluten? Do you recycle? Did you inhale? What do you think about Harry and Meghan? How about the Kardashians?

It's crazy, right? Or, maybe you haven't noticed all the divisions and separations. To be honest, not noticing would be perfectly valid. Even (or perhaps especially) in church. We don't notice because, so often, our congregations are filled with people who look like us, dress like us, talk like us, work like us, were raised like us, eat like us, sing like us, vote like us, etc.

One of the most visual examples of our self-imposed imaginary dividing lines in churches is race. As much as we wish things were different, our churches are still largely segregated by race and ethnicity. To use my own denomination as an example, a 2014 Pew Research study revealed that, while American Baptist Churches USA is categorized as a "more diverse" denomination, 73 percent of individuals who identified themselves as ABCUSA were white. To be fair, this statistic is somewhat skewed because studies rarely allow respondents to indicate multiple religious affiliations, and many ABCUSA-aligned African American congregations are dually aligned with other Baptist conventions, including historically black Baptist groups.[1] In contrast, the American Baptist Churches Information System (ABCIS) report shows more than 52 percent of ABCUSA members are African American and 36 percent are Euro American, with other racial and ethnic groups each comprising between 1–4 percent of the denomination's membership.[2] Yet, while ABCUSA membership may be racially and ethnically diverse, our individual churches are not. Only 301 of 5,030 ABCUSA congregations self-identify as multiracial.[3] So, even though, if one were to gather all American Baptists members in one location, we would be richly diverse, on any given Sunday morning, we remain deeply segregated.

Whether a congregation is white, African American, Latinx, Haitian, or Karen, the racial-ethnic lines have been drawn. Though we interact with people who look different from us at work and school and at community businesses, we fail to notice the lines we've drawn around ourselves on Sunday morning. Or, we notice but accept it and let it be. Because it works.

Until it doesn't.

On August 12, 2017, in Charlottesville, Virginia, Rev. Dr. Matthew Tennant found himself positioned between a skinhead and a counter-protester. He was, essentially, a bystander at the Unite the Right rally. Eventually, the rally erupted in chaos; one person was killed, and Dr. Tennant was left with questions to ponder. This book is the result of that experience.

Maybe you were there that day? Perhaps you saw the images on television. Maybe you have been involved in other marches or rallies or protests or riots, whether as participant or passer-by. Maybe you've sat at Thanksgiving dinner, locked in disagreement with a family member over the Second Amendment or tariffs on China or whether a professional football player should kneel during the national anthem. Maybe you've drawn an invisible line to divide the backseat so your sibling will keep her distance on a road trip from Michigan to Virginia.

Wherever you've drawn your lines, this book invites you to revisit them. You may realize, for the very first time, that you have drawn lines in the first place. You may decide you still need them. You may still want them, right where they are. But you might decide those lines have been there long enough and think to yourself, "I'd like to figure out how to cross over that line." This book, my friend, is a good place to begin your journey.

—Deidra Riggs
Author, ONE: *Unity in a Divided World*
Bloomfield, Connecticut
January 2020

Notes

1. See https://www.pewresearch.org/fact-tank/2015/07/27/the-most-and-least-racially-diverse-u-s-religious-groups/. In the same 2014 Pew study, 53% of African Americans reported affiliation with historically black denominations, as opposed to 4% as mainline Protestant, in which ABCUSA is included.

2. Data provided courtesy of the American Baptist Churches Information System (ABCIS) (January 17, 2020).

3. Internal reporting by ABCIS indicates 56% of our churches are Euro American, in contrast with roughly 25% being black (African American and Haitian). Another 3% are Asian Pacific congregations, 7% are Hispanic, and less than 1% are Native American, with some 3% identifying as "other." Data provided courtesy of the American Baptist Churches Information System (January 17, 2020).

Acknowledgments

John Donne said, "No man is an island, entire of itself; every man is a piece of the continent [sic]." I am not an island, and I did not develop this project alone. After an event hosted by the Baptist News Global called "Conversations That Matter" in 2017, where about twenty-five ministers gathered in Cambridge, Massachusetts, to discuss ministry in a polarized and politicized U.S., I began thinking about what to say in this divided culture. The generosity of the Baugh Foundation, who sponsored the event, played an important role in the initiation of this project, and for that I am grateful.

I am indebted to the many authors cited in the pages of this book. I am equally indebted to many authors not cited in this book, but who influenced my ideas. To the congregation of Hickory Baptist Church in Whitakers, North Carolina, thank you for taking the risk and calling me to my first pastorate. To the congregation of Kilmarnock Baptist Church in Kilmarnock, Virginia, thank you for allowing me to serve you for eight grace-filled years. To the congregation of University Baptist Church in Charlottesville, Virginia, thank you for the current blessings and opportunities and for being a prophetic voice in a divided world.

To Lynn Martin, thank you for your detailed comments on my manuscript.

To everyone at Judson Press, thank you for your partnership and commitment to bringing relevant works to Christians today. Rebecca Irwin-Diehl, thank you for working with me again and for sharing your editorial talent. Lisa Blair, thank you for your skill in facilitating communication among team members and for keeping momentum on the project. Wendy Ronga, thank you for designing a cover that matches what I was trying to say. To the rest of the Judson team, thank you.

To my wife, Melanie, you are my best friend and a great help—this project is stronger because of you. Thank you.

While I am grateful to everyone who offered suggestions or advice, any mistakes that remain are entirely my own.

Introduction

On August 12, 2017, I was near the "Unite the Right" rally in Charlottesville, Virginia. By *near*, I mean that I stood in the street, walked along the sidewalk, and remained close to the protesters and counter-protesters. I did not participate in the protest or the counter-protest. Rather, I was a bystander who neither participated in resisting nor fought against the hatred. In some ways, I felt invisible. People focused their energy on each other. They fought, threw objects, punched, and shoved one another. I walked my bike in the same direction as the neo-Nazis and skinhead marchers. In that moment, I needed to cross the street, but the marchers stopped momentarily to wait for the group ahead of them to move. The crowd stopped moving too. Ahead of them, a skinhead yelled profanities, and a counter-protester yelled profanities back. They were face to face. There was no alternate route, so I walked my bike in between the skinhead and counter-protester yelling at each other. I walked right between them. They paused for a moment. Then, once I passed, they resumed yelling.

In that pause, I saw both division and our shared humanity. They were each a human being. If someone cut their skin, they would bleed. If someone tickled them, they would laugh, like Shakespeare's Shylock said.[1] I do not know what the two men saw. I do not know how I appeared to them. I imagine that, like Holly Golightly in *Breakfast at Tiffany's*, they had the "mean reds." It is different from "the blues." It is a feeling rooted in anger and causes a momentary lack of awareness of the surrounding world.

She explains: "The blues are because you're getting fat and maybe it's been raining too long. You're just sad, that's all. The mean reds are horrible. Suddenly you're afraid and you don't know what you're afraid of."[2]

In the pause, I saw shared humanity. Maybe the two men were afraid, and perhaps they did not even know what scared them. But they stopped yelling as I walked between them. Maybe they stopped yelling as they recognized the humanity that they shared with me. Recognizing another person can disrupt the "mean reds" and make a person aware of the world again. As soon as the interruption ends, fears can return.

Fear is a driving force for many people. People want tighter borders, because they are afraid of immigrants. People want lower taxes, because they fear the government misusing their money. People want tariffs, because they fear losing their jobs. People want a strong military, because they fear foreign powers. People want less regulation, because they fear not having the freedom to operate their business as they choose. People want rules, because they fear not knowing when they are right or wrong. Despite some apparent incongruities, fear drives each idea.

Looking to the past, we can find examples of living with fear. Giving in to fear could have changed the world. To better understand the barriers, fear, ourselves, and our world, we can turn to various Bible passages and put them in conversation with the politicized world of the early twenty-first century. Consider Jesus. Or, rather, consider the birth of Jesus. Luke sets the scene in Nazareth of Galilee. Mary is engaged to Joseph in Nazareth of Galilee. She is a virgin, and for the miraculous conception of Jesus Christ, this is an important detail. She is a young Jewish woman and educated in the ways of her faith. She knew what we would now call the Old Testament, but to her, they were simply the Scriptures.[3]

We often spiritualize Mary's encounter with the angel and her role in bringing God Incarnate into the world. Yet, when the angel appears to her, Mary is not an empty vessel to which we might ascribe meaning. She had a life up to this moment—a life that prepared her for her response. This is the same with all of us. It is the same for the two shouting men in Charlottesville. Our lives prepare us for the way we respond to various situations. How does Mary respond? How did her life prepare her for her calling? Was she afraid?

The angel says, "Greetings, favored one! The Lord is with you" (Luke 1:28, NRSV). What does it mean to be "favored" by God? The Bible says Mary was "troubled" (NIV), "confused and disturbed" (NLT), and "perplexed" (NASB). I translate dietarachth as "wholly disturbed." The angel reassures her and tells her that she will bear the son of God. Would we be perplexed if God sent an angel to greet us? Would the two men stop their conflict if an angel appeared? Would they be perplexed?

What about us? How have our experiences prepared us for encountering God? Mary had her lifetime of experience to prepare her for her encounter with the angel. Even though she was young and her experience was limited, it taught her what to expect. She was engaged and could look forward to having children at some future point. The Annunciation ran counter to her experiences.

When we consider our experiences, what do they teach us about overcoming differences? What do our experiences teach us about speaking a unifying word to deep divisions? Or, have our experiences left us unprepared? Are we more influenced by our culture than by God? Are we forming tribes of like-minded people? Are we fed by the algorithms of our social media feeds, where every message complements our history of "likes" and "shares"? More and more, culture seems to feed division. If I disagree with something, I can unsubscribe to it. If I do not like what a friend says, I can stop following his or her posts. Social media are not unique. The same herd mentality bleeds into real life.

Churches divide over political and ideological differences. There are few, if any, places where divergent viewpoints are both welcomed and encouraged. Does a big tent that welcomes all ideas mean that people do what is right in their minds? Not necessarily. People can adhere to different political ideas while remaining in contact with one another. They do not need to divide over each difference.

This epidemic of division infects people across perspectives. Conservatives make lists of rules and identify litmus tests. If you disagree with one of them, you will not be included in the tribe. Liberals do the same thing. For example, consider same-sex marriage.[4] Broadly speaking, conservatives are clear about their opposition to same-sex marriage, whereas liberals accept it. However, neither side has much tolerance for someone who holds a divergent perspective. If someone holds to most other conservative positions, yet accepts same-sex marriage, her fellow conservatives will not be tolerant. Before liberals pass judgment, they should recognize the planks in their own eyes. Liberals are often quite intolerant of someone who holds most other liberal positions but opposes same-sex marriage.

My goal is to reverse a world of either/or examples and present a scriptural response to polarization. The result of this reversal can help people recognize and honor our shared humanity. This recognition undermines a herd mentality and enriches the experience of sharing our humanity with one another. When we spot what we share with other people, we can empathize with them. We can see the world from their perspective. Even if we do not suddenly agree with them, we can value their humanity. Constructive engagement with one another can help overcome barriers.

What about Mary? Did she live in a world of either/or situations? In Luke, Mary faced barriers to being the mother of the Lord. She was young. She was female in a male-dominant world. She faced losing her social safety net. Yet God favored her. Again, what does it mean? Why is she perplexed? And does being favored necessarily bring fear? It does not sound like something scary. The favorite employee gets preferential treatment. The teacher is lenient with the favored pupil, maybe even giving an extra curve while grading. Being favored sounds like something good.

For God, being favored means being given a special task. Mary was no exception. God had a special task for her. This was a blessing that Christians see as integral to our tradition. But what a strange blessing! God choosing Mary did not raise images of blessing to mind—she did not get rich, advance her career, or have more friends from this blessing. There were also potential consequences from a virgin pregnancy—there was a chance Joseph would not believe her. He might have left. Her parents might not believe her either. They could have cast her out. Then, she would be a pregnant teenager at a time when the social safety net consisted of family. Again, we might think, what a strange blessing! Instead of experiencing the blessing of being favored, she experienced the heartbreak of her son's execution as a seditious criminal.

Being favored by God seems as if it should include special benefits, especially as defined in contemporary culture. Being favored by God should improve our social standing, wealth, and good health. But, as Alan Culpepper writes, "Acceptability, prosperity, and comfort have never been the essence of God's blessing."[5] Instead of allowing culture to define partiality, we can look at what Scripture, tradition, and theology suggest being favored would mean. God favors everyone. The idea that God favors everyone helps set the stage for overcoming the barriers that separate us from one another.

Being favored by God means experiencing God, knowing the author of life and creator of the universe. Being favored by God is what leads people to spend time in Bible study or to sacrifice themselves in service. Faith experiences enrich people's lives, and sometimes we can see it only after the fact.

Mary, God's favored one, experienced a strange blessing. Her story fits with the ancient "birth of the hero" myth. In this myth, a woman, preferably one of noble origin, has no children until quite an advanced age. She experiences a divine revelation announcing the birth of a son. Sometimes the vision alludes to the son's fate. The ensuing response can vary. But in all cases, in due course, the child is born and manages to fulfill his destiny despite dangers which would have conquered a lesser mortal.[6] In this sense of the "birth of the hero" myth, Mary is favored. Jesus, God incarnate, arrives. He overcomes obstacles and fulfills his mission.

With Mary, there is something more significant than Luke pushing a "birth of the hero" myth. Luke's Mary is special. She models faith. She is a model for people today to accept Jesus. Mary plays a significant role in the greater narrative of Luke. She is the first to believe. She is the first to tell others what she has heard. She is the first disciple.[7] For Luke, being a disciple means hearing the word of God and incorporating it into one's life. Mary heard God speak through the angel and responded by incorporating this word into her life.[8] In this sense, she is a model for every disciple who follows, including us. This experience makes her very much a favored one—one who experiences the favor of God's overshadowing presence.

Although Mary's experience of being favored has been the focus, the Bible repeats the idea that God is with humanity and transcends human boundaries. Mary's story speaks to our story as well. God is with us. As we explore overcoming divisions and polarization, we cannot forget that God being with us fulfills God's purposes, not our goals or a cultural definition of fulfillment or satisfaction. Yet, even committed Christians miss God's purposes. Here are some of the ways we get it wrong:

Jesus says, "Blessed are the poor," and we praise the rich.

Jesus says, "Blessed are those who mourn," and we avoid sadness and suffering people.

Jesus says, "Blessed are the meek," but we feed the egos of the proud.

Jesus says, "Blessed are those who hunger and thirst for righteousness," and we celebrate those who hunger and thirst for success.

Jesus says, "Blessed are the merciful," yet we look for retribution and revenge.

Jesus says, "Blessed are the pure in heart," but we overlook infidelity, lying, cheating, jealousy, gossip, meanness, cruelty, and hurtful acts.

Jesus says, "Blessed are the peacemakers," and we create wars and conflict.

Jesus says, "Blessed are those who are persecuted for righteousness' sake," and we retreat from bearing witness to our faith.

The Bible goes far beyond the Beatitudes.

Jesus says feed the hungry, give clothing to those who need it, and visit sick people and those who are in prison (Matthew 25:31-46), but we have starvation, poverty, and loneliness.

Scripture says, "God loves all people equally" (Galatians 3:28), yet we create divisions.

James says faith without works is dead (2:14), yet we provide no actions to accompany God's calling.

What do faith and actions look like? What does it look like to live out God's calling? Culpepper writes, "If Mary embodies the scandal, she also exemplifies the obedience that should flow from blessing."[9] She embodies what it means to have God with us. She symbolizes being favored by God. Is peace easy? Is overcoming fear easy? No. Being favored includes a different kind of good. It is not like shallow Facebook posts like, "My husband made breakfast. #blessed." Or, "My daughter did the dishes. #blessed." Peace and overcoming fear require hard work. Overcoming divisions does too.

During the "Unite the Right" rally, my participation was minimal. I did not do any hard work. I did not understand what was happening. I sat on my bike and watched. My family and I had moved to Charlottesville two weeks before the rally. Before the rally, the church where I serve as pastor held a service praying for peace. Praying is what I know how to do.

> If there is a mass shooting, we pray.
> If there is a natural disaster, we pray.
> If there something else bad, we pray.

Before the rally, it made sense to pray for peace.

That morning, when I arrived downtown, I saw a group of my new minister-friends standing with locked arms, blocking the entrance to Emancipation Park. I did not stand with my fellow clergy, locking arms, because I did not feel a sense of God calling me to confront the neo-Nazis. Instead, I smiled and waved at the one or two people I recognized. As tempers broiled into violence, I wanted to shout, "Hang on! Let's pause for a second. What's going on? Shouldn't we be praying?"

Being a bystander is not hard work, and praying is not enough. Mary was not a bystander. The Bible does not list the bystanders in her story. God calls people to do more than stand around watching and organizing the occasional prayer vigil. My desire to pause reflects some of the frustrations I experience when I do not know what to do.

What does the alt-right really believe? What do the counter protestors believe? The ministers whom I know believe in making peace and spreading love. But some protesters, who supposedly also espouse peace and love, shouted profanities and threats of violence. Some people carried weapons. Some people were clearly looking for a confrontation. In a context like the rally/protest in Charlottesville, there is no space for dialogue. It is tense. People yell. They shout slogans. The other side screams back. People scuffle, bunch up, and punch and push each other. As long as we remain in the crowd, shouting and taking sides, we cannot hear one another's stories, learn the other's perspective, or recognize God's presence in the midst of the tension.

When tensions rise, people shut down. Pain can accompany fear and foster divisions. It is like Holly Golightly's "mean reds." When people experience the "mean reds," they cannot engage in dialogue. They cannot empathize with the other person's feelings or experiences. Having a conversation to discuss differences can produce results. How can we see past our "mean reds" and find the shared humanity in other people?

The following pages are organized into four parts: (1) the quest, (2) the journey, (3) life, and (4) getting there. Each part moves along the lines we draw in partisan politics and explores the questions: How can we overcome differences? What does God say to our desire to find a tribe and isolate ourselves from outsiders? Is it possible to overcome polarization by crossing the lines and moving toward harmony?

Notes

1. Cf. William Shakespeare, *The Merchant of Venice*, ed. Jonathan Bate and Eric Rasmussen (New York: Modern Library, 2010), Act 3, Scene i.

2. Blake Edwards, "Breakfast at Tiffany's" (Paramount Pictures, 1961).

3. M. Pauline W. Lewela, "Mary's faith-model of our own: a reflection," *AFER* 27, no. 2 (1985).

4. For the sake of simplicity, this example will follow a binary conservative/liberal orientation and ignore potential nuances.

5. R. Alan Culpepper, "Luke," in *New Interpreter's Bible*, ed. Leander Keck (Nashville: Abingdon, 1995), 52–53.

6. Athalya Brenner, "Female social behaviour: two descriptive patterns within the 'birth of the hero' paradigm," *Vetus testamentum* 36, no. 3 (1986).

7. Raymond Edward Brown, "The Annunciation to Mary, the Visitation, and the Magnificat (Luke 1:26-56)," *Worship* 62, no. 3 (1988).

8. John Burns, *Modeling Mary in Christian Discipleship* (Valley Forge, PA: Judson, 2007).

9. Culpepper, "Luke," 53.

‖‖‖‖‖‖‖‖‖‖‖‖‖‖‖‖‖‖‖‖‖‖‖‖‖‖‖‖‖‖‖‖‖‖‖‖‖‖‖

PART ONE

The Quest

In 2002, at the Strictly Sail Chicago Trade Show, I led a seminar on navigation at sea. To navigate successfully, one must know a number of things, including current location and directional heading, speed, location, and distance in one hour, and the new location. Each step is important, and each step is part of the navigational process. Writing everything down is important. Having good charts and reliable instruments is important. However, the key to navigation is knowing where one wants to go. Everything else is secondary. To navigate, one must know where one wants to go. It seems simple, but many marine accidents begin with someone setting off without a clear understanding of where they want to go and how to get there.

In a polarized and politicized culture, knowing where to go and how to get there remains a challenge. Where do people of faith want to go? Unity. A place where all people matter. But, what does unity look like? Does unity mean everyone agrees on everything all the time? That sounds boring. Does unity come only with a shared enemy, as in World War II? That sounds violent. *E pluribus unum* is the unifying motto of the U.S.—*out of many, one*. The motto is the opposite of violence. People come together. Through their differences, they grow stronger. They learn from one another, not in an idealized utopia, but through growth.

Is achieving unity possible? What about harmony? Harmony is the pleasing combination of different notes, sounding simultaneously. Harmony produces chords and chord progressions that makes music sound good. Harmony requires different notes. In a polarized society, individual people can remain true to their own convictions while coming together. Differences can create something beautiful. Perhaps harmony is a better goal than unity.

Harmony allows people to have a dialogue while disagreeing with one another. How can U.S. society move in that direction? Is it through a process of transformation? If so, what does it look like? Are there easy steps to follow? No. There are certainly no easy steps. If there were, then more people would follow them. Then, how do we begin? Scripture. The Bible is an ancient collection of writings from many different hands in various places. So, we begin defining our quest with the last book in the Bible: Revelation.

Why begin at the end? In the navigation analogy, knowing where one is going is essential for getting there. Starting in Revelation can help us define our quest and identify where we are trying to go. Revelation 21:1 talks about a new heaven and a new earth. It also talks about a time when the "sea was no more." Instead of navigation at sea, we navigate humanity. Revelation uses images to convey something about God and God's relationship to humanity.

Revelation 21 offers a striking promise. Consider the historicity of the chapter. When Revelation says, "I saw the holy city, the new Jerusalem, coming down out of heaven from God…" God's revelation has nothing to do with the modern nation-state of Israel, which could be a polarizing interpretation of the biblical notion of Israel. Yet, in our milieu, the United Nations created Israel after World War II—less than a lifetime ago. The "holy city" or New Jerusalem of Revelation 21:1 is about God bringing humanity into harmony. It is symbolic and looks to the future with hope. Different people come together. With all of their uniqueness, the "holy city" brings people into rich chords of God's music. Divisions in the U.S. in the early twenty-first century pale in comparison. Pablo Richard writes about the idea of Jerusalem in Revelation:

> In Revelation, Jerusalem is a myth, a symbol for the people of God or the community... History is not limited to being a society of men and women, but also includes the cosmos or nature. The new cosmos and the new humankind are bodily. Transcendence... means overcoming death, chaos, and darkness, not overcoming bodily-ness and history. In the new world created by God there is bodily-ness and social relations, but they are now without death, chaos, darkness, and oppression.[1]

If we want to get to a place where all people matter, or a place where we can have dialogue, then we can explore shared experiences. The "holy city" of Revelation is a place of transcendence. Our quest is finding common ground with others. The "other" includes people with whom we disagree. This quest is not isolated, but in communion with others and in communion with God. The quest functions in the new cosmos and includes transcendent presence of God. God understands humanity. Humanity never fully understands God because God is beyond human comprehension.

The holy city provides some idea of where the quest can go. This vision of the New Jerusalem in Revelation expands the divisions to be overcome. The New Jerusalem is a new world and goes beyond political differences. Instead of limiting overcoming polarization to Democrats getting along with Republicans, we realize that nature and the cosmos surround everything. Overcoming political differences is meaningless if we destroy the planet. Harmony means working with nature, not subduing it. Revelation adds new dimensions to what we have to overcome.

Too often, people see Revelation as a rosy description of heaven.[2] Instead, it is crisis literature, produced during a time of great difficulty. Revelation was most likely written at the end of the first century, after the fall of Jerusalem. Ancient Israel was vastly different from modern Israel. It was usually occupied from the time of Solomon going forward. For example, in 66 CE, a Jewish uprising occupied Jerusalem and kicked out the Romans. In 70 CE, the Romans besieged and conquered Jerusalem. Josephus describes the destruction of Jerusalem: "the army had no more people to slay or to plunder, because there remained none to be the objects of their fury."[3]

In Revelation, Jerusalem is a symbol. "I saw the holy city" means God is here. It means "I have a dream." Martin Luther King Jr.'s famous speech echoes the hope present in Revelation. God is present. Revelation offers strength and hope in desperate times. This might seem unlikely based on some of its stranger visions and dispensational interpretations. Revelation is an apt message of hope for difficult times.

The challenges and difficulties of today are real to the people who face them. Regardless of whether the twenty-first century is an easier place to live than the first century, which it arguably is, people face challenges today. A teenager who complains about being ignored feels ignored. An employee who says she is undervalued feels undervalued. Someone who says he has no friends feels like he has no friends. Whether someone agrees with the legitimacy of the feeling or not makes no difference. If that teenager's sense of dislocation goes unchecked, add some violent tendencies, abuse, bullying, or mental illness, and the recipe for a school shooting is complete. The same logic applies in various situations. In that moment, it does not matter whether life in the twenty-first century is easier than in the first.

From Revelation to King's famous speech, hope is present in the world. Each person experiences hope for the future in different ways. Each person has a distinct perspective. Only you know the view from where you stand. What do you see? Or, rather, what is the worst thing in the world? Is it death? Revelation 21 assures people who face death that death is not the worst thing. God destroys death! Over and over in the New Testament, God reigns over death. Jesus raised Lazarus. Peter raised Tabitha. Paul raised Eutychus. Easter celebrates God's dominion over death. Death is not the end.

The quest, then, is not to overcome death. God has already accomplished that. Is the quest harmony? Yes, but what does harmony mean? Revelation 21 explores a "holy city." Is the holy city one that is full of harmony? Or, does it represent the eschatological hope of the future? Resituating our present notions is part of the process of overcoming the barriers that separate us from other people. People do not need to attempt something that is impossible, like overcoming death, because God accomplishes the impossible. The quest is to have a desire to overcome the barriers that separate people from one another.

Part of the quest is reconciling our relationship with God's grace. Grace is the free gift that God offers to humanity. The new heaven reflects that grace. These visions of grace in Revelation are about a world in which God's idea of reality becomes our idea of reality. The process of growing in our relationship with God puts us in a metaphorical new Jerusalem. This vision of grace is not just a vision for the future. Travis Poling writes: "I am not convinced that the new Jerusalem is only a vision of what the end may look like. I believe it is also what the world looks like every single time our ways of seeing and believing and loving and living in this world come to an abrupt end, and God comes in to transform our lives. In this sense, the world ends over and over again."[4]

To get to the destination, even though we know the end, we must know where we are now. It is the present. Poling's description of transformation places the new Jerusalem of Revelation in our path.

Instead of looking back in time to an ancient or imagined history, we can look forward. In this sense, end does not mean end. It means rebirth. The world ends in the sense of the old self ending and the new self, God's regeneration, begins. Revelation teaches about hope for the present, not merely about the past troubles or a future apocalypse.

It is easy to frame Christianity as something, someday, about the sweet by and by. Faith is about this moment, now, what each of us does here, in this life, today. When God transforms people's lives, they enter a radical new world of faith. The world ends and a new one begins.

Where are we going? The quest begins with faith. What does it mean? How does faith impact polarization in the U.S.? Preparing for the quest means listening to God. Before moving to the journey, we can look at the nature of faith.

Faith involves risk and risk-in-Christ. The quest is harmony. Harmony is not unity, because each person retains her or his individuality. Together, people can be greater than the sum of their individual parts. To achieve harmony in Christ, people reconcile with God. And, each faith step involves risk. Addressing divisions is a risk. Confronting evil is a risk. Speaking truth to power is a risk. These risks are worthy steps in faith toward the quest.

Notes

1. Pablo Richard, *Apocalypse: A People's Commentary on The Book of Revelation*, trans. Phillip Berryman (Maryknoll, NY: Orbis, 1995), 161.

2. Gordon D. Fee, "(Revelation 21:1—22:5)," The (Original) Tale of Two Cities, Part 2: God Makes All Things New," in *Revelation*, A New Covenant Commentary (Lutterworth Press, 2011), 289ff.

3. Flavius Josephus, *The Works of Josephus*, trans. William Whiston (Peabody, MA: Hedrickson, 1987).

4. Travis Edward Turner Poling, "Every time the world ends: John 21:1-25, Revelation 21:1-4, 22:1-5," *Brethren Life and Thought* 51, no. 4 (2006): 240.

1

Growing Faith

"The kingdom of heaven is like a mustard seed that
someone took and sowed in his field; it is the smallest of
all the seeds, but when it has grown it is the greatest of
shrubs and becomes a tree, so that the birds of the air
come and make nests in its branches."

—Matthew 13:31-32

"Faith is taking the first step, even when you don't see
the whole staircase."[1]

—Martin Luther King Jr.

Every step we take can be part of our faith journey. Everything we
do can be a building block of faith. What is faith in this context?
Does it mean something risky, like a leap of faith? Martin Luther
King Jr.'s description of steps on a staircase with a mysterious
future implies faith as risk-taking. Does faith connect with belief,
as in matters of faith? The parable of the mustard seed suggests
growing belief. Is it relational and part of a community of faith?
For overcoming differences, all three notions of faith enter into the
conversation. Risks, belief, and relationships become part of the
faith journey at different points along the way.

When we go through our daily routine, whatever we do, we have
the opportunity to invite God on the journey. Everything fits into

the faith journey even if it does not feel like it. We cannot see around the twists and turns ahead. When King gave a speech at New York's Park-Sheraton Hotel in 1962, he could not know the path ahead of him. It was hard and he would face challenges, physical pain, emotion turmoil, and major setbacks. During this speech, he issued a call and challenge in one sentence, "Our lives begin to end the day we become silent about things that matter." When we have faith, we continue the journey.

In each step of our journey, we can reflect God's grace in the way we interact with others. We can smile, say something kind, or see the world through another person's eyes. These are simple things, and they may not amount to much. But they have no cost and are a place to start. When we carry out these simple acts, we share God's love. That love plants a seed. That seed can grow into something big. As it grows, these seeds of faith and God's love heal divisions. Jesus' words about the mustard seed reflect the way small things can grow into something big.

For Christians, healing divisions involves faith. People cannot overcome differences apart from the faith they hold that things can be better. This is faith in the sense of belief or hope for the future. Without faith, there is no reason to heal divisions. Without faith, especially a cooperative faith that reflects a communion of believers, there is no reason to come together. In a world where people choose sides and then surround themselves with like-minded people and insulate themselves from any challenging or divergent thoughts, all aspects of faith can die.

The idea of seeds of faith healing divisions might sound simplistic. The seeds do not grow on their own. Small things are fragile by themselves. If our entire faith rested on smiles, kind words, and empathy, we would have a fragile faith. Fortunately, faith builds and grows. It also experiences setbacks and can contract. Each step brings us closer or farther away from God. During some seasons of life, we grow. During others, we do not. At those times, God seems far away.

A Distant God

What do we do when God feels far away? Some years ago, during Sunday morning worship, I announced that I needed to leave immediately after worship to drive to a service installing my friend in his new pastorate. After church, a member offered to fly me to the service in his private plane. During the flight, he brought up President Obama and said, "He disappoints me almost every day."

The whole conversation was a bit surreal. We were in an airplane, and this man was giving his time to me. He was giving his talent. He was giving the fuel and operational costs of his airplane. He was profoundly generous—flying me to my friend's installation service. Yet, in that moment, he saw a completely different world than I did. While I might not agree with all of Obama's policies, he did not seem disappointing. He maintained decorum and, as a politician, seemed to treat people with respect. What did this man see that I did not see? How was Obama so disappointing? I did not ask. The best pastoral response can appear to be disengagement. Or, was I supposed to speak the "truth in love" (Ephesians 4:15)? I could say, "What is so disappointing?"

Was that space open? Had we established a level of trust in which we could have an honest conversation? I struggle with remaining silent, yet I want to keep the option for further conversation open. I am concerned that speaking the truth even in love can end the opportunity for future conversations.

Even if I asked, could he have articulated his disappointment in the safety of the cockpit of his airplane? Was it simply that Obama's politics are liberal and this man is conservative? This distinction represents an ideological difference, not the basis for disappointment. Did race play a part? I do not know. For the man who gave of his time, talent, and airplane fuel, God was missing in Obama's America.

The psalmist cries out, "Save me, O God, for the waters have come up to my neck" (Psalm 69:1). There are times when God appears to be distant or absent. Life can swell, like rising waters, and it feels as though there is no hope. The desperation of Psalm 69 reminds contemporary readers that it is okay to recognize when we feel that God is distant. For the man flying me to my friend's installation service, God had abandoned him to an Obama-led America. What words can address polarized perspectives? Part of addressing polarization is seeing the world through another person's eyes. To see through another person's eyes means adopting a pastoral spirit across partisan lines.

> *Keep faith.*
> *God will be back.*
> *God is in control.*
> *We can get through this.*

Would words like these be helpful? These words might provide opportunities to interact with people who share some basic tenets of faith. Or, are they empty platitudes that provide little relief when it feels like God is distant? U.S. Presidents come and go. A long view of history might provide some perspective. In the immediate present, each step can contribute to knowing God better and growing stronger on our journey. Paul Tillich refers to God as the "ground of our being"[2] —the basis of everything we do. When Christians see God as the "ground of being," Jesus' parables of faith provide the fertile soil for spiritual growth. Faith grows on or in something.

Fragile acts like kind words and empathy come together with the other actions and experiences. They form the tapestry of our faith journey. Together, they grow and our lives become a reflection of God's love and light. It continues throughout life. Regardless of whether someone is a Christian from the cradle roll or a brand-new believer, the faith journey continues throughout life. We can all always grow in Christ and learn something new.

Too often, we hold an image of God as high, exalted, transcendent, this inexpressible thing that is beyond our comprehension. Karl Barth referred to God as "a Wholly Other majestic and unobservable unity…"[3] On the one hand, Barth is right about not being able to describe God. On the other hand, he understands the movement of the Holy Spirit. He understands that God is active in everyday life. He understands that faith is a journey. This is a journey where God still speaks and can relate to humanity in a variety of ways. Barth's thought developed over his career, and he writes about the ways God can speak. Following Jesus' example of using everyday items in Matthew 13, Barth uses some common references—things his readers would be familiar with and might find a bit shocking. This was the 1930s, and he wrote, "God may speak to us through Russian Communism, a flute concerto, a blossoming shrub, or a dead dog. We do well to listen…if [God] really does."[4]

Imagine God speaking through these things. A blossoming shrub or a flute concerto is believable. Russian communism or a dead dog? Barth does not assume God speaks through these things, but he recognizes the impossibility of containing God. With a transcendent God who moves and speaks in everyday life, we do well to listen… *in case God really does!*

In the Gospel of Matthew, Jesus spends time with people—he talks and shares everyday life with them. In Matthew 13, he tries to help his listeners understand God. He says, "The kingdom of heaven is like…" Then, he talks about things with which the people are completely familiar. He uses mundane examples five times for this "kingdom of heaven is like…" metaphor. These examples bridge the gap between Jesus and his listeners. Just as Barth's readers were familiar with his examples (i.e., a flute concerto, Russian communism, and a dead dog), Jesus' listeners were familiar with his examples.

Faith Like Yeast and Seeds

Jesus' listeners knew about the mustard shrub. Jesus speaks to their active imagination. They could look past him and probably see a mustard shrub, or as soon as he mentioned it, they could all picture it. Through familiar elements and our active imaginations, we can reintroduce a distant God. Jesus' listeners knew the small size of mustard seeds, they knew how the shrub grew, and they knew how big it was. Likewise, they knew about yeast. In Matthew 13:33, Jesus says, "The kingdom of heaven is like yeast that a woman took and mixed in with three measures of flour until all of it was leavened." They knew how to get yeast: let a small portion of old bread sit; let it spoil. If they do not wait long enough, it is useless. If they wait long enough, yeast forms and they can make bread. If they wait too long, it ruins the bread and will make people sick.

Faith grows like the yeast in the flour. Faith grows like a tiny seed. Like yeast and seeds, conditions and timing can either foster or disrupt growth. Bringing harmony to a polarized and politicized America also requires care, timing, and the right conditions. If faith is part of the Christian life, then a growing faith will be part of the process of overcoming differences. Like preparing yeast or planting a seed, people must carefully consider faith and can expect it to produce transformation in people's lives.

In Matthew, we find a sower, weeds, wheat, mustard seeds, and yeast. In Barth, there is Russian communism, a concerto, a shrub, and even a dead dog. God *can* speak through anything. Examples of ways God can speak surround each person, every day. Some examples come from experience or looking at the world. Jesus spoke in parables, using familiar items to explain the unexplainable. When I am hiking in the woods, the beauty of nature reminds me of God at work in the world. When I see someone helping another person at The Haven, a local shelter for homeless people, I see living parables.

If we look at the parables of life around us, we can see God at work, and our faith will grow. In these experiences, we can find a shared humanity. Faith is recognizing God at work and seeing how God is speaking through various things in the world. The kingdom of heaven parables in Matthew 13 show what the world can be like. The mustard seed grows into something big (Matthew 13:31-32). Yeast makes flour expand (Matthew 13:33). The treasure in a field is a source of great joy (Matthew 13:44), and the pearls are a great value (Matthew 13:45-46). The net brings everything together, and then the good and the bad are separated (Matthew 13:47-48). The parables show peace and reconciliation and a harmonized world. They show faith blossoming and taking root. Jesus tells his listeners what God's kingdom is like. *It starts small and grows big.* This goes against the grain of popular assumptions about what is important.

Powerful or Meek?

In twenty-first-century U.S., we celebrate big and powerful things and people. Celebrities and celebrity-politicians frequently talk about wealth and success. Yet hubris (pride in human achievement that overlooks divine favor) is antithetical to the way of Christ. It is the opposite of who God wants people to be.

In a culture seeking wealth and success, powerful people cast aside meek and mild people. The small things, like the metaphorical mustard seed and yeast, are disposable. We want to hear from the rich and powerful, people with high positions. If I told someone, "My buddy Paul once said, 'Creation is a space filled with love and joy,'" she might think, "that's nice" or, "Who is that guy to say something about creation?" But, if I said, "Paul Fiddes, Professor of Divinity for the University of Oxford, wrote in *Seeing the World and Knowing God*, 'Creation is a space filled with love and joy,'"[5] she might take notice and say, "What did Prof. Fiddes mean?"

Even in positive ways, we ascribe credibility to position. Instead of assuming credibility based on the prestige of the source, we can approach each and every interaction positively.

Returning to the idea of God's presence, Fiddes makes a helpful suggestion that creation is not a space from which God is absent. Instead, creation is where people give and receive in the name of love. In that space, God is present. Like seeing God moving in the world, "Creation is a space filled with love and joy" and the presence of God. Jesus' parables in Matthew 13 point to the potential that is present in the world. Fiddes is exploring Job 28 and the sense of place in God. Where is the place of wisdom? Wisdom cannot be found in a literal place at all. It is in a "no-place."

This idea of "no-place" uses postmodern theologies to situate the inner movement of God-self. Specifically, Jacques Derrida's deconstructionism challenges attempts to establish an ultimate meaning.[6] Since God is transcendent, God does not occupy a particular place. This does not mean God does exist, but God is in a "no-place" that is beyond a particular place. In the current landscape of U.S. politicization, polarization presents binary positions that fit in a particular position. The quest to overcome this division requires a risk-taking faith based on the faith of belief and grows within a community of faith.

Power, wealth, and status are uncomfortable with the "no-place" of God's wisdom and the idea that love and joy are not for sale. When everything in life is commodified, nothing has any value. The seed and the yeast turn power on its head. Matthew 13:31-33 is nestled between the parable of the weeds among the wheat and its explanation. Each parable subverts the conventions and norms we associate with God and the world. Small things matter, not the great and powerful.

Faith is small. It grows. It matters. And, it connects to overcoming differences.

A Nicaraguan Response

Wisdom can come from a variety of sources. It is easy to look to scholars for insights. It makes sense. Cultural conditioning points us to experts. But what about average people? The Nicaraguan priest Ernesto Cardenal was saying Mass among the impoverished people of the Solentiname Islands. He became convinced that God could (and would!) speak through the people listening to Mass. So, he read the Gospel lesson, set up a tape recorder, and asked people what they thought. He transcribed, edited, and published the tapes in his four-volume work *The Gospel in Solentiname*. The lessons speak to crossing lines—the line between clergy and laity and trained and untrained.

When the Jesus of the New Testament speaks in familiar words, using familiar experiences, he invites us to join him. As people consider the faith journey, we can use familiar, everyday experiences to express our faith. Consider Marcelino. Marcelino is one of those souls known only to God and, perhaps, a few hundred people in his life. In all likelihood, he was a poor, possibly illiterate fishing person from the Solentiname Islands on the southern end of Lake Nicaragua. Here is how he describes Matthew 13:31-33: "I don't know about the mustard seed, but I do know about the *guasima* seed, which is tiny. I'm looking at the *guasima* tree over there. It's very large, and the birds come to it too. I say to myself: that's what we are, this little community, a *guasima* seed."[7]

Marcelino might not be a great theologian like Karl Barth or Paul Fiddes, but he captures the essence of Jesus' message. "I'm looking at the *guasima* tree over there." Each person can ask, "What do I see? What is familiar that I can use to express my faith?" One of Marcelino's friends, Manuel, describes the faith journey as follows: "At first it seems insignificant… but afterwards it grows… this tree is the transformation of the world."[8]

The tree of faith can transform the world, and each person gets to be part of transformation. Regardless of where we are on our spiritual journey, our experiences, or the divisions we have encountered, faith is that little grain that can take root and change our lives and the lives of those around us. One person's faith nurtures another's. As the journey continues, "this tree is the transformation of the world."

Notes

1. Martin Luther King, Jr., "Speech at the Park-Sheraton Hotel" (New York City, New York State Museum, September 12, 1962).

2. Paul Tillich, *Systematic Theology, Volume 1: Reason and Revelation, Being and God* (Chicago: The University of Chicago Press, 1951), 110.

3. Karl Barth, *The Epistle to the Romans* (London: Oxford University Press, 1933), 452.

4. Karl Barth, *The Doctrine of the Word of God*, ed. G. W. Bromiley and Thomas F. Torrance, trans. G. W. Bromiley, vol. I/1, Church Dogmatics (Edinburgh: T & T Clark, 1936/1975), 55.

5. Paul Fiddes, *Seeing the World and Knowing God: Hebrew Wisdom and Christian Doctrine in a Late-Modern Context* (Oxford: Oxford University Press, 2013), 264.

6. Cf. Jacques Derrida, *Of Grammatology* (Baltimore: Johns Hopkins University Press, 1976).

7. Ernesto Cardenal, *The Gospel in Solentiname, Volume 2*, trans. Donald D. Walsh (Maryknoll, NY: Orbis, 1978), 54.

8. Cardenal, *Solentiname, Vol. 2*, 51.

2
What Is Faith?

Faith is a fine invention
For Gentlemen who *see!*
But Microscopes are prudent
In an Emergency![1]
—Emily Dickinson

In 1830, a prominent family from Amherst, Massachusetts, gave birth to a sociable little girl named Emily Dickinson. Most likely she led a normal childhood, but as she grew older, she became increasingly withdrawn. The world now remembers her as the reclusive poet who rarely left her family's home.

Dickinson uses this poem to articulate a worldview where both science and religion can fit together. Using scientific language, such as "invention" and "microscopes," she sees science as a tool to better understand religion. The poem represents harmony between two ostensibly, opposing worldviews. By using whatever tools we have at our disposal, we can see God more clearly.

Faith leads to action, trust, and reliance on God. Faith is seeing God and growing in a divine relationship. Faith is participatory. Each person does not rest on a past event, like having gone to Sunday school as a child. Faith is the active desire to grow closer to God and others in the community.

One way we grow closer to others is to pray for them. By lifting up another person in prayer, it becomes much more difficult to hate or embrace divisions. Beyond defining prayer, the simple act of uttering someone's name before God connects the one praying and the one named in prayer. Regardless of what happens in prayer, speaking to God about someone who sees the world in a different way can break down barriers.

Through prayers, does the other person experience broken barriers too? Maybe. Maybe not. You cannot control other people or the experiences they have when you pray. People can only control their own actions. Starting with the quest for unity and recognizing the role faith plays in this journey is part of the process of overcoming divisions. "Faith is a fine invention" that can undermine polarization.

In this sense, sacred words feed the faith journey. Like Barth's Russian communism, flute concerto, blossoming shrub, or dead dog, God can speak through anything. Those things through which God speaks are holy expressions. Understanding the definition of faith is an early step on this journey and a way to open oneself to God's voice.

What is faith? Is it *just* an invention? Faith is complete trust in something or someone. Christian faith means wrestling with God. Dickinson does not imply anything diminutive in her poem. Calling faith an invention does not diminish its usefulness in responding to the hate and division in the twenty-first-century United States. Hebrews 11:1 says, "Faith is the assurance of things hoped for, the conviction of things not seen." Reconciliation is something that is unseen and happens at a future point. It is an act of faith. To pursue reconciliation assumes that we begin from a point of separation. To seek reconciliation in the U.S. means acknowledging the starting point of not only polarization but separation. Beginning with separation, faith is a useful tool or invention for looking ahead and seeing the possibility of overcoming differences.

Faith means Democrats and Republicans can get along. What about people from different ideological worldviews? Can they come together? Again, faith's answer is, "Yes!" Coming together does not mean surrendering all beliefs and giving up everything that makes one an individual. It means addressing differences and finding common ground. It means trying to get along. To make any attempt to move forward is to live in faith.

America First?

Faith is positive. It hopes *for* something. It looks *to* the future and imagines something better. In a perfect world, everyone wins. The gospel has a universal message: "God so loved the world" (John 3:16-17). God reconciles everything back to God-self (Colossians 1:20). In the gospel, there are no winners and losers in the sense of winners and losers in games. God loves people from the U.S. just as much as people from other countries, like Mexico.

During his inauguration speech, the newly-elected President Donald Trump said, "From this moment on, it's going to be only America first. Every decision on trade, on taxes, on immigration, on foreign affairs will be made to benefit American workers and American families."[2] With this declaration, finding common ground for mutual understanding can be challenging. From his words, he seems to lack a sense of mutuality. But, what does "American first" mean? What does it mean to him? What does it mean to the United States? How might God view the idea of "America first"? Is it possible to overcome the polarization produced by such a dichotomous worldview? "America first" runs counter to faith's positive approach for the future. The country needs reconciliation.

In Matthew 20:1-16, Jesus compares the kingdom of heaven with a landowner who went out early in the morning to hire people to work in a vineyard. He goes back to the marketplace throughout the day, hiring workers each time. At the end of the day, the workers expect some sort of tiered pay scale. Instead, the landowner pays them all the same wage.

This parable is part of a larger scene beginning with the story of the rich young ruler in Matthew 19:16. Matthew borrows material from Mark 10:17-31 and adds a bit from the Q source. If contemporary readers want to imagine the story as Jesus told it, they must resist seeing the landowner as an allegory for God and the payment as the last judgment. Setting aside this allegorical leap, the human tendency would be to identify with the workers hired first.

The parable about the laborers in the vineyard transforms winners and losers into a communal gathering. Each person receives what they need. Is the parable about salvation? Yes. Faith connects with every aspect of life. If faith in Christ were only about salvation, then Christians could set aside faith matters in daily living. One could do as one pleases and trust that God has no ethical or behavioral expectations on a Christian's life. In the eyes of God, we could do whatever feels good. Scripture does not support this hedonistic interpretation. Instead, the Bible is explicit: faith connects with action (James 2:14-17). God has expectations of God's people (Micah 6:8).

What about the laborers in the vineyard? Were the people who showed up to work all day losers because they did not earn a greater wage than those who arrived at the end of the day? No. There was enough for everyone. What about this "America first" mantra? Salvation is for all, not the winners. Life is for everyone, not just Americans.

A few chapters after the parable about the laborers in Matthew, Jesus tells a Pharisee about the greatest commandment. In Matthew 22:37-40, Jesus says, "'You shall love the Lord your God with all your heart, and with all your soul, and with all your mind.' This is the greatest and first commandment. And the second is like it: 'You shall love your neighbor as yourself.' On these two commandments hang all the law and the prophets."

Jesus' comment, "hang all the law," adds gravitas to a loving statement. He does not describe some namby-pamby *love*. This love is powerful. It transcends boundaries, even borders. Elsewhere he answers the question, "Who is my neighbor?" In the Gospel of Luke, the neighbor is a *foreigner!* A Samaritan, of all people! Thus, I ask, how is a Mexican different from a Samaritan, if, in this analogy, Americans are the chosen ones of Israel? "America first" misses the mutuality of the Christian faith. It falls far short of "love your neighbor as yourself."

Listening

Jesus reacts to people and situations. He listens. He hears what they are saying and how they experience the world. But the Bible is full of other characters who listen, take in the message, process it, and respond with varying degrees of effectiveness or grace. Some biblical heroes reflect humanity. Since Jesus was both human and divine, the humanity of Jesus differs from the humanity of other people in Scripture. Unlike Jesus, they are flawed and we can learn from the good and bad in their lives. These characters can help Christians grow in faith by demonstrating the way humanity relates to God.

In Genesis, the patriarch Jacob is an unlikely hero. He teaches lessons about listening, processing, reacting, and reconciling. Many aspects of his story suggest that he is not a hero. He tricked his father, cheated his brother, profited at his father-in-law's expense, and demanded a divine blessing. Yet, he is one of the founders of our faith. In some ways, Jacob's life is a reminder that God can use anyone to do anything. When trying to find models for reconciliation in the face of a polarized world, Jacob points to a God who moves, speaks, and acts according to God's purposes. Sometimes those purposes can confound our minds, but usually faith, hope, and reconciliation are not far beneath the surface.

Jacob was Isaac's second son and appears to be an unlikely source for the fulfillment of God's promises. He was Abraham's grandson. He was the first of the patriarchs to have many children. Through Jacob, God's promise (Genesis 15:5) to Abraham appears possible. Abraham and Sarah only had Isaac, and they had him when they were old. Several times, God says to Abraham, "I will make you a great nation." Up to this point, fulfillment seemed unlikely. Abraham's only sons were Isaac and Ishmael, and he sent Ishmael away. Isaac fathered Esau and Jacob. Esau was the first-born and a rugged individual. Jacob, according to Genesis 25:29, liked to cook. He used his prowess in the kitchen to cheat Esau and get Isaac to bless him. The promise now appears possible because Jacob has so many children. Unlikely sources often teach valuable lessons about God.

Ernest Hemingway wrote in a 1935 *Esquire* article, "When people talk, listen completely. Don't be thinking about what you're going to say. Most people never listen. Nor do they observe."[3] Listening is a skill. Jacob used his listening skills to betray Esau. He knew Esau would trade his blessing for some stew because he listened.

Listening does not always lead to something bad, that is, to manipulation or exploitation. But, in Jacob's case, listening helped him understand Esau. He knew what to do and how to trick his brother. He knew that his brother could be shortsighted and impulsive. After a hard day of working, a meal might seem like an equal trade for their father's blessing.

Listening enables us to understand the other person. When we stop what we are doing and focus our attention on the other person, we can learn something. To learn something new, we must be fully present with the other person. As Hemingway said, "Most people never listen." Too many conversations consist of politely waiting for the other person to shut up so that we can speak. Listening means engaging with the words the other person uses.

Listening applies to the way people relate to God. Too often, we find our own solutions to our problems. Even if Christians pray about a problem, solutions and actions accompany a request for God's endorsement rather than humbly seeking God's direction. Rather than allowing God to direct us, we too often look for a quick solution. Finding God's direction means listening completely for God's voice.

God Speaks

God speaks, whether people are ready to listen or not. Perhaps one message from Jacob's life is to see God present and working in the world. When Jacob cheated Esau out of his birthright (Genesis 27:1-29), it seems wrong. Yet God continued working in spite of Jacob cheating his brother. Gerhard von Rad warns against trying to find *the* meaning of this story.[4] But, perhaps our meaning is to keep looking, keep seeking, keep trusting— because God is here, active, and a participant in the stories of our lives. Jacob is no worse or better than any of us. If listening helps him to engage with God, then we can listen to engage with God. When he dreams of a conversation with the Lord in Genesis 28, he bolts upright and proclaims, "Surely the LORD is in this place."

God still speaks. We need to listen. Instead of hearing God, we become distracted. We fixate on issues—political, social, economic. We focus on the divisions. She is with *that* party. He is a *single-issue* voter. In either case, the position belongs to the person who holds it. Despite being convinced that I am right and others are wrong, I am no closer to God's perfection than anyone else. When God is speaking, my views (or yours) is do not matter. God's voice is the one that matters.

Our responses to issues should be the product of hearing and experiencing God. Our views should not influence what we think we hear God say. The Christian ideal is to keep God primary and make the issues secondary.

What happens when we engage with God? God returns the favor. God engages with us. Can we take it further? What happens when we grab hold of God? Does God grab hold of us? Maybe the onus is on us. In Genesis 32, Jacob is getting ready to reconcile with Esau. During this process, he engages with God—he grabs hold of God.

Sometimes people resist God's voice. God offers peace, joy, and love. This division between human desires and what God offers is the source of many of the problems in our world today. People think they have the answers, and human hubris knows no bounds. Many people view humility and contentment as weakness. In the movie *Wall Street*, Gordon Gekko says, "Greed…is good."[5] But it is not. Love, joy, and peace are good.

In Genesis 32, when Jacob is set to make peace with Esau, he experiences terror. Esau is powerful and Jacob is terrified of him. In the night before their encounter, Jacob sent his wives, maids, eleven children, and all of his things ahead across a ford in the Jabbok Stream. When he was left alone, the Bible says "a human" shows up and wrestles with him until daybreak. It uses the Hebrew word *en-oshe'*, which is different from the more dignified *aw-dawm'*. Both mean *person* or *mortal*.

Genesis 32:24 is tricky because a few verses later, it says that Jacob wrestled with God. The implication seems to be that God showed up in human form and physically wrestled with Jacob, leaving him with a hip injury. Or, could God have been working through a person? Some innocent bystander who happened upon Jacob after he sent his family and possessions ahead? When the prophet Hosea references this event (Hosea 12:4), it refers to Jacob having wrestled with a messenger or angel.

Does Genesis 32:24 suggest that when we engage with a person or another mortal, we are engaging with God? It certainly supports the ancient notion of *imago Dei*. Humanity is made in the image of God. It also points to God's activity in and through the world. If God is moving and active, then God can move and act through people. Thus, Jacob's wrestling match is with God, whether he physically grabs hold of God or a mortal.

Wrestling with God

The devious Jacob, the one who tricked his brother (Genesis 27) and worked for fourteen years to get his bride (Genesis 29), has come a long way. He engages with God. He reaches out and grabs the Lord. Whether his grappling was physical or he had a vision while in a deep state of prayer or he was dreaming, he is not unscathed by his encounter with the divine. He emerges from this encounter a changed person. Grappling with God has the same impact on people today.

In this passage, the human or angel saw that victory was impossible, and therefore knocked Jacob's hip out of socket—yet the two continued wrestling. Jacob was fully engaged. When his opponent saw the sun was about to come up, he said, "Let me go." But Jacob, the trickster who just a few chapters ago was willing to go to deceptive lengths—including dressing up like his twin brother Esau to get his father's blessing—insists that this person bless him. The messenger renames Jacob *Israel* in Hebrew—one who strives with God.

How does wrestling with God impact our effort to overcome polarization and politicization? When people are gripped by God differences and divisions can fade. Politics, social issues, economic and environmental injustice—these are important issues of our time. As we consider them, we cannot let go of God. God is present with us as we gather next to the Jabbok Stream.

We can grab hold of God. We can wrestle with God. When we do, we do not enter a physical contest with the divine. We engage. We can wrestle with God in the depths of our souls and demand answers. The same book that talks about God creating the world (Genesis 1–2) does not show God incapable of defeating Jacob in a wrestling match. This passage does not say Jacob was winning. The two were entangled and had no place to go. Jacob could not get what he wanted and the person, or *en-oshe'*, could not get what he wanted. As Frederick Buechner writes:

> God is the enemy whom Jacob fought there by the river...and whom in one way or another we all of us fight—God, the beloved enemy. Our enemy because, before giving us everything, [God] demands of us everything; before giving us life, [God] demands our lives—our selves, our wills, our treasure.
>
> Will we give them, you and I? I do not know. Only remember the last glimpse we have of Jacob, limping home against the great conflagration of the dawn. Remember Jesus of Nazareth, staggering on broken feet out of the tomb toward the resurrection, bearing...the proud insignia of the defeat that is victory...[6]

Faith is wrestling with God. Faith is trusting that God will wrestle back. Faith is looking into the empty tomb and trusting that Jesus will come out. Faith is belief in the future, trusting that something good *can* happen. Faith is not blind hope that something good *will* happen. It trusts in God's assurance that everything is in God's hands. Political divisions have always existed. During some periods, the chasm between different sides is bigger than at other times. Christians can have faith that God is bigger than our differences.

Notes

1. Emily Dickinson, *Emily Dickinson Collected Poems* (Philadelphia, PA: Courage Books, 1991), 113.

2. Donald Trump, "The Inaugural Address," (2017), https://www.white house.gov/briefings-statements/the-inaugural-address/.

3. Ernest Hemingway, "Monologue to the Maestro: A High Seas Letter," *Esquire*, October, 1935.

4. Gerhard von Rad, *Genesis: A Commentary*, Old Testament Library, (Philadelphia: Westminster Press, 1961), 314.

5. Oliver Stone, "Wall Street," (1987).

6. Frederick Buechner, *Secrets in the Dark: A Life in Sermons* (HarperCollins, 2007).

3
Taking Risks

"Peter stepped down from the boat, and walked on the waters toward Jesus. But, when he saw that the wind was strong, he was afraid and beginning to sink, he cried out, 'Lord, save me!'"

—Matthew 14:29-30

"There is no way to remove the moral risk of human action."[1]

—Juan Luis Segundo

Overcoming anything involves risk. To play a game means risking defeat. To learn in a class means risking failure. To take a job means risk. To love another person means risking hurt. Everything involves risk. Addressing politicization is full of risk. Overcoming it carries repeated risks due to its long and arduous process.

The story of Peter and Jesus walking on water in Matthew provides some tools for dealing with the risks of the faith journey. Jesus told the disciples to get into a boat. He dismissed the crowd. According to Matthew, it had been a busy period of ministry. He had been teaching, fed five thousand, and reached a time for a well-earned break. The disciples set sail, while Jesus retreated to a mountaintop for some long overdue solitude.

Then a storm blew in. The wind and the waves took all of the disciples' attention. The same is true of encountering storms on the water today. When the wind picks up, and the waves start crashing into a boat, the deck sways back and forth. The boat creaks and makes unnatural noise. The boat can ride down a wave, accelerating before crashing into the next wave with a jarring deceleration.

Storms take all of one's concentration. Problems that seemed large on shore do not exist in a storm. The only thing that matters is the storm. The storm distracts us from everything else. Likewise, an opposing viewpoint can function as a storm, distracting us from the bigger picture. Instead of seeing the other person, we hear an abhorrent notion of seeing the world. That notion becomes an object. We fixate on the point at which we disagree.

Single-issue voters may have the most trouble seeing a broader picture. The one issue takes precedence over all others. God calls people to focus on Christ. Then, in watching Christ, we can learn to take a broader perspective and love people where they are. This does not mean that we keep them where they are or discourage them from changing perspective when matters of justice and peace and love are at stake. It also does not mean that we keep our own narrow focus. Perhaps, in dialogue, we might change too.

Talking with someone can overcome divisions. It has that potential. When churches create opportunities for dialogue, they are at their best. This is a positive version of what the church can be in the world. It runs counter to the cultural expectation of the church as separate from the world. Reconciling in the face of division is how Christians can be in the world but not of it (Romans 12:2).

What are the risks? Certainly there are risks involved with entering into dialogue with someone who holds a different position. The other person might not be open to dialogue. Reconciliation means acknowledging separation. Two friends who know they hold to different political parties can ignore the difference and remain friends. They never deal with their differing worldviews. Is this a healthy way to maintain a relationship? To a certain extent, yes.

But it is a disingenuous relationship because the two people are not honest about their beliefs. Having faith in the possibility of overcoming separation might deepen their friendship. What is at stake? Being honest risks their friendship. Before beginning the journey of reconciliation, recognizing the risk is a first step.

For Peter, following Jesus involved risk. It meant stepping out of the boat. This story is only in Matthew and Mark. Matthew adds some details to the Markan version and cleans up some of the confusing geography. Luke, who had access to Mark, does not include this story. Most likely, even first-century readers had trouble accepting the miraculous walking-on-water event.

Post-Resurrection Reading of Appearances

Like the other three Gospels, Matthew's account was written after the resurrection. We read the words today as if they are happening in real time. Yet, they are a memory, seen through post-Easter lenses. Instead of tripping over the miracle, as Luke's early readers might have done, we seek the meaning, recognize the risks involved, and ask how it applies to the politicized climate of twenty-first-century United States.

Our modern minds hear the story of Jesus walking on water and question how he could defy the law of gravity. A century ago, readers inspired by Rudolf Bultmann sought to demythologize the miracle.[2] They might say, Jesus was on a reef or in shallow water. But God does not worry about the physical laws. Easter symbolizes God's power over death. If God has power over death, then walking on water should not be a problem. Thus, the question to bring to this story is theological, not scientific. What does the story say about God? What can it say to people who are struggling with political divisions?

God overcomes chaos. The Greek word for walking, in this verse (Matthew 14.29), is *per-ee-pat-eh-o*, meaning "to tread all around" or "walk at large." M. Eugene Boring interprets it to mean "conquest." The sea is a symbol of the "active power that threatens the goodness of life." He writes, "To be at sea evokes images of death."[3] Walking on the water symbolizes conquering death. The Hebrew background of Matthew points to Jesus' post-resurrection power to save people.[4] Matthew tells about Jesus walking on water to show Jesus' power over nature. In turn, the story points to his power over death.

Seeing Jesus frightens his disciples. Although some scholarly treatments of this miracle highlight Greco-Roman mythological parallels, Jesus' miracle was novel and difficult for Matthew's initial audience.[5] The story is about the authority of Jesus. He walks (conquers) on the sea (death). This larger section in Matthew (13:53–17:27) is the formation of the new community. Jesus is pulling away from public life. His relationship with the disciples deepens. At the same time, his opposition grows. He continues to challenge the crowds, but they neither reject him nor become disciples.

Relating Risks to God

Reading about a pre-crucifixion appearance after the resurrection means suspending knowledge of what happens next. In that moment, Peter is following Jesus but has no way of truly knowing where Jesus is going. He might believe Jesus is the Messiah, but he cannot fully understand what it means to be the Messiah. When Jesus invites him to walk on water, he could be taking his final step. This is the risk he carries when he steps out of the boat. Jesus has not conquered death yet.

We live life in real time. Each day, each step is the present. Like the frightened disciples in the boat, we can look at the tumult of waves and seek safety. We avoid the risk. Even if Jesus calls us, our tendency is to stay in the relative safety of the boat. Earlier, Jesus told them to "go on ahead to the other side" (Matthew 14:22). The wind pushed the disciples away from the shore. In their present moment, they are in a storm. When they saw Jesus, they were terrified. Some thought he was a ghost. Others cried out in fear. In that moment, fear and uncertainty overwhelmed their ability to try to make sense of what was happening.

Setting the fear aside, even for seasoned sailors like the disciples, storms are hard work. When the deck sways to and fro, it takes energy just to remain in one place. First, the sailor tenses muscles to keep from having her legs buckle. Then, she relaxes to keep from getting knocked over. Alternating flexing and relaxing, tensing and letting go—this goes on throughout a storm. For the disciples, it goes on all night.

Sailors have an old saying: "We cannot change the wind, but we can adjust our sails." The disciples might not have liked the storm. Some might have been afraid, but they were probably handling it. They were adjusting and working. The Bible does not say the disciples were afraid of the storm. It says, "...the wind was against them." Literally, the wind was *en-an-tee-os* or "contrary." Only when they see Jesus walking on the water are they afraid.

What is it about Jesus that is frightening? In the story, fear comes from the surprise of seeing a person where there should not be anyone. The fear comes from defying physics. Jesus walks on water. But this fear is metaphorical too. Jesus was treading on stormy seas. Instead of retreating to the safe shores of our own tribes, Jesus calls humanity to join together in the midst of the storms of life. Ephesians 4:1-6 says,

> I therefore, the prisoner in the Lord, beg you to walk worthily of the calling with which you were called, and with lowliness and humility, with patience, bearing with one another in love; being eager to keep the unity of the Spirit in the bond of peace. There is one body, and one

Spirit, even as you also were called in one hope of your calling; one Lord, one faith, one baptism, one God and Father of all, who is over all, and through all, and in us all.

This is a call for unity. It is not alone. The Bible repeats this call over and over again. When Jesus appears and challenges us, provoking us to move beyond our comfort zones, Jesus can be frightening.

For the disciples in the boat, the question is who is this that walks on water? Most English translations render the Greek, "It is I." But Jesus says, *eg-o i-mee*, or, "I am." This might not sound like much to us, but Jesus quotes the Septuagint, or Greek translation, of Exodus 3:14. Moses asks, "What shall I say is the name of the one who sent me?" God says, "Tell them I am (YHWH) sent you." This does not mean that Matthew conflates Jesus and YHWH. Recall, Jesus had just been praying to YHWH. Instead, this passage is about God's presence as mediated by Jesus and Jesus as present in the community of faith. God is with us.

Whatever risks we face, God is there. Better yet, we can see through the trajectory of God's interaction with humanity. God has been there, continues to be here, and will be wherever we are in the future. The entire scriptural tradition points to God's continuing presence. Polarization and politicization in the U.S. might seem new, but it is not. It is not new in the U.S.—recall the Vietnam era or the Civil War—nor is it new in history. God continues to move.

Leaving the Safety of Our Boats

Leaving the boat can be the hardest step. Even though a boat can be a scary place in a storm, it is far more secure than the water. Other than Coast Guard rescue swimmers, who would jump from a sound vessel into the crashing seas? In the early morning hours, the disciples' energy was almost spent. In the glimmer of the light of sunrise, Peter yells above the wind and waves, "Lord, let me come to you" (Matthew 14:28).

Jesus says, "Come."

I want to argue that Peter's leap is not getting in out of the boat but in leaving the community. Traditionally, we envision Peter as the model for responding to God and taking risks. Instead, Peter represents the disciples. The boat is the community of faith. Peter leaves the community, and, when he is alone, he sees the violence of the storm. Then, he begins to sink.

Sometimes, taking risks for God means staying in the community and moving together in faith. The real risk is not stepping out of the boat—it is leaving the community. When the violence of the storms of life crashes all around, standing alone is almost impossible. When Peter notices the waves, he becomes afraid and starts to sink. He shouts, "Save me, Jesus!" Jesus takes Peter's hand and leads him back to the boat. In taking his hand, he leads Peter back into the community. Then, the wind ceases.

What are the risks God is calling you to take? The answer will be as unique as each person. Each person can answer individually. However, you do not have to leave the boat to answer. We can remain in our faith community, in our churches. Together, Christians can listen to what God is saying.

Overcoming differences, polarization, and politicization means taking risks. Relationships are at risk. There is a risk of being isolated and excluded. Strong differences elicit strong opinions. Closely held beliefs bring emotional responses. We can face the risks because God goes with us. God does not send each person out alone. When Peter leaves the boat alone, Jesus takes his hand, directs him back to the boat, and then joins him there. Likewise, God is with us in the boat of Christian fellowship, walking alongside us as we seek unity with other people.

Notes

1. Juan Luis Segundo, *Liberación de la teología* (Buenos Aires: Carlos Lohlé, 1975), 123. Cf. Juan Luis Segundo, *Liberation of Theology*, trans. John Drury (Maryknoll, NY: Orbis, 1976), 109.

2. Rudolf Bultmann, *New Testament and Mythology and Other Basic Writings*, trans. Schubert M. Ogden (Philadelphia: Fortress Press, 1984).

3. M. Eugene Boring, "Matthew," in *New Interpreter's Bible*, ed. Leander Keck (Nashville: Abingdon, 1995), 328.

4. Patrick J Madden, *Jesus' Walking on the Sea: An Investigation of the Origin of the Narrative Account*, vol. 81 (Walter de Gruyter GmbH & Co KG, 2014), 6.

5. D. McPhee Brian, "Walk, Don't Run: Jesus's Water Walking Is Unparalleled in Greco-Roman Mythology," *Journal of Biblical Literature* 135, no. 4 (2016): 763.

PART TWO

The Journey

The journey of faith can take different twists and turns along the way. We never know what to expect along the way. Psalm 139:7-12 assures people of God's presence always. No matter where people go, God is there. Although the psalm has a geographic tone (e.g., "If I go to the other side of the sea, you are there"), I interpret it to include God's temporal presence. God is with us along all of the twists and turns of the journey, going ahead of us in time and seeing where our path may lead.

We do not know the future and sometimes it brings wonderful surprises, like meeting new friends or having new opportunities. A seemingly random conversation might lead to a new connection. As a minister, I speak with people every day. Some conversations are part of the way life seems to plod along—meeting with a contractor, discussing budget items, or planning some church logistics. The beauty of each conversation is the part that I do not know. I never know how a conversation will go. What starts as a normal conversation can twist and turn and become a rich, meaningful connection.

One day, I stopped in the church kitchen. A volunteer was getting ready for a community meal and the other volunteers had not arrived yet. I stopped just to say hello and tell him I appreciated the time he was giving to the church. Then I asked him about his daughter. She is in her thirties, married, and has a child. He and his wife told me about a terrible incident involving their daughter and how it fractured their family. What could have been a mundane conversation, like any other, changed. As we talked, I felt God's presence. "Even there [in the kitchen of the church] your hand shall lead me" (Psalm 139:10).

The journey of life moves through different phases, and in each one, other people are part of it. In chapter four, we will look at the other people on the journey. Who do we include? Who do we exclude? How do we decide? The people along the way impact our journey. The man in the kitchen influenced my day. I felt bad for his experience of brokenness. I thought about each member of his family and the way the situation affected their lives. I prayed for them. I also listened and was present when he wanted to talk. I hope that my presence, prayer, and words positively impacted his day.

The man in the kitchen continued to impact my day after I left. I thought about my children and prayed for them. I thought about other people I know who experience pain and shattered relationships. I prayed for them. It made me think about a friend who is struggling with alcoholism. I sent him a text, and he replied that he had been sober for seventeen days. I congratulated him.

The impact of the other people on my journey that day continued to have a ripple effect. Someone touched me and it reminded me of God's presence. In chapter five, we will look at that impact. Just as the man in the kitchen prompted me to move in a positive way, people can have a negative effect on one another. People can bring one another down. People struggling with addiction can experience a trigger event and quickly lose the power to stand against their addiction.

Paul connected growing in faith with worship and saw worship and the Christian life as full-bodied experiences. For him, worship is integral to the journey. Worship is the chance to gather with other people who also want to grow in faith. It becomes a building block for overcoming differences. Along the path, Christians represent Christ when with friends, at the store, at school or work, running errands, or at church. The faith journey with other people impacts relationships, the way they spend their time and money, and the way they think.

To be in a relationship with another person involves trust. In chapter six, we will explore trust and truth-telling. What is the truth? How can we know that the psalmist was accurate when we read, "If I ascend to heaven, you are there" (Psalm 139:8)? How trusting can we be? What if, instead of building a connection with another person and fostering trust, we violate it? The relationship will break down or die. At that point, overcoming divisions starts to feel impossible. There is less material to work with. Still, finding something to use to build trust is worth the effort because people are worth more together than apart.

Once I saw a friend on the other side of the street. She was waving when I looked up and saw her. It was a busy city street and we were between two corners. I waved and was tempted to wait for a break in traffic to cross the street and catch up with her. She pointed to the nearest corner, and we each walked to the corner where there was a crosswalk. When the light changed, I crossed to her side and we had a nice conversation. There was nothing spectacular or special about it. It was normal. *What have you been up to? How's it going?*

Not every encounter will be transformative. Not every step on the journey will be noteworthy. But, on the path, we go with God. In chapter seven, we will look at that path. We will explore the parts we do not see. We cross streets to seek God and continue the journey.

4
Eradicate or Include

"Knowing that we can be loved exactly as we are gives us all the best opportunity for growing into the healthiest of people."[1]

—Fred Rogers

"For God so loved the world that he gave his only Son, so that everyone who believes in him may not perish but may have eternal life."

—John 3:16

Each person must decide with whom she or he will travel on the journey. We choose our traveling partners. Do we push people away? Or, do we draw people in? God loves us exactly as we are. If Mr. Rogers is right, God also loves others as they are. John 3:16 echoes this sentiment. Hitler, however, identified an entire group of people whom he tried to exclude. These extreme examples paint a stark picture. It appears that a person can decide whether to be inclusive or exclusive. Yet, is the world clearly divided into two groups—those who include and those who exclude? No. Every person can choose the good. Every person can choose the bad.

In the stark political climate of twenty-first-century America, we ascribe an identity to other people. The ones from our tribe or group are good; others are bad. This either/or approach paints a dichotomy of the other person as for or against us. Constructive relationships with people who feel differently about an issue or look different become challenging. This dichotomous approach to others comes from all directions—including some surprising sources.

One surprising perpetuator of a for-or-against-us worldview was heard in the inaugural address of the U.S. President in 2016. During his inauguration speech, Donald Trump said, "We will reinforce old alliances and form new ones—and unite the civilized world against radical Islamic terrorism, which we will eradicate completely from the face of the earth."[2]

The inauguration speech for any president is an opportunity to unify the country after an election. Every election has partisanship. The 2016 election was particularly bitter especially since Trump won the electoral college but not the popular vote and because Russians meddled in the election. These factors contribute to the partisan divide. Thus, it would seem that the newly elected president would want to bring the country together. Yet, a phrase like "eradicate completely" is the opposite.

Words matter. To "eradicate" means to destroy completely. To *eradicate poverty* means *to put an end to poverty*. Thus, to "eradicate completely" means total destruction of "radical Islamic terrorism."

When people talk about terrorism, they must choose their words carefully. One cannot deny that terrorists exist, but it is also the case—a crucial point—that the majority of Muslims are neither radical nor terrorists. (In fact, non-radical Muslims have suffered at the hands of terrorists.) Terrorism is a big issue and requires a multinational approach, which the president alluded to ("reinforce old alliances and form new ones"). Unfortunately, when the U.S. attacks terrorism by bombing villages, these attacks inspire some people to join the terrorists. For example, a drone strike on a village where terrorists are hiding might lead other villagers to become sympathetic to the terrorists. Terrorists succeed when they invoke terror. All of us should choose our words carefully. Carelessly tossing off words like "eradicate" to sound tough might inspire some people, but when taken seriously, "eradicate" harkens to darker times like the Holocaust. So, how can the U.S. respond to terrorists? How can Christians respond to the fear caused by threats of terrorism and demonstrate love for one's neighbor in this context? The first two responses require multinational cooperation.

First, countries and NGOs can identify the causes of terrorism: poverty, inequality, cultural changes, and religious fanaticism. When people address poverty, inequality, and cultural changes, religious fanaticism has more trouble recruiting adherents. When there are no choices, people might be more open to the possibility of joining terrorists. How can we address poverty? Individuals can support efforts to address the causes of terrorism by buying fair trade coffee and bananas and other products. Christians can support programs that engage in economic development and education for poor and impoverished people. For example, Kiva (www.kiva.org) is a microfinance company that allows people to lend money to low-income entrepreneurs in over 80 countries. Supporting people through organizations like Kiva undermine people's motivation to join terrorist organizations. And we can ease away from "America first" ideologies, as we discussed in previous chapters.

Addressing the causes of terrorism is like reading the parable of the Good Samaritan (Luke 10:25-37) and trying to figure out why the robbers attacked the man going from Jerusalem to Jericho. It is the question behind the question. Instead of killing one angry hornet, treat the nest. Terrorists do not fall from the sky, fully-formed with a suicide vest strapped on. They have a lifetime of experiences that make joining a terrorist organization seem like a viable path forward. In the Luke passage, what caused the robbers to start robbing?

If we want to predict where terrorists will come from, one indication is poverty. Fanatics like Osama bin Laden or Abu Bakr al-Baghdadi might come from wealth and espouse terrorism as a means to achieve a particular ideology. But, the soldiers—the ones who carry out the terrorist attacks—almost universally come from poor backgrounds. Identifying areas of extreme poverty can be a leading indicator of potential terrorism.

Economic disparity is also a leading indicator. In Palestine and Israel, poverty falls along the border between the two lands. Witnessing this difference probably frustrates Palestinians. It does not excuse terrorism, but it does help explain why someone might become violent and blame the wealthy West. Cultural shifts and lack of opportunity compound the problems.

Second, to fight terrorism, the U.S. can develop partnerships with countries who are already engaged in fighting terrorism. Regional partnerships would be more effective than drones and bombs. I do not know the intricacies of defeating Daesh (aka ISIS), but a thoughtful response would lead to greater success than bombing the entire area or attacking terrorists' families, as Trump suggested. Backed by a U.S.-led coalition, Iraq pushed Daesh out of Mosul in July 2017. Working with local partners led to victory.

Third, fear is the currency of terrorists. We must not be afraid. Big talk, like "eradicate," reveals inner fear. Instead of playing into the hands of terrorists, we must rise above it. The Bible assures us that God controls the world. If Christians believe it, then we have nothing to fear. In 1 Chronicles 28:20a, David says to Solomon, "Be strong and of good courage, and act. Do not be afraid or dismayed; for the Lord God, my God, is with you." We are not alone. This is good news. As we face polarizing politicization, God is with us. When we prayerfully seek ways to reconcile differences, God is with us. God is even with us when we try to figure out how to deal with terrorism. Humanity is not alone. We can live out our faith in this world, even when it gets scary, and God remains with us.

How can Christians demonstrate love for one another in a context that includes terrorists? We can address the causes of terrorism. We can seek to be agents of reconciliation. We can listen to one another, trying to understand the conditions that led the robbers to attack the man going from Jerusalem to Jericho. We can look for the question behind the question. We can seek to include other people and speak out against inflammatory rhetoric like "eradicate completely." Words matter. Let us engage in thoughtful discourse about complicated problems and avoid vacuous platitudes. When we start locally, by relating to our neighbors, we can gradually widen the circle and change the world—even if we can only change part of it.

The Problem of Evil and Creating Boundaries

How do we address the problem of evil in our world? Phrases like "eradicate completely" have an evil ring to them when applied to an entire people group (e.g., the Holocaust). When dangerous words become the norm, they impact real people. Then, people who have nothing to do with partisan politics can have their world turned upside down. Bad things happen, and Rabbi Harold Kushner's question comes up: "Why does God let bad things happen to good people?"[3] The Russian philosopher Nicolas Berdyaev answers this question by attributing most evil to human freedom.

God gave us freedom. The question is what to do with it. We use it to do good things—help people, feed the hungry, exhibit hospitality, visit the sick, and work for justice. Berdyaev writes, "Freedom is the fatal gift which dooms humanity to perdition."[4] Why does freedom lead to eternal punishment? Because we can be judgmental and closed-minded, say hurtful things, and much, much worse. But we have a choice. It is through obedience to God that we grow in faith. When we grow in faith, we might not know the answer to the question "why does evil exist," but we develop words to discuss it.

These words provide language we use to address the existence of evil. People say things like, "It is what it is," "God's got a purpose for you," or "It was part of God's plan." Every iteration of each phrase is an attempt to talk about evil. Christians in the Reformed tradition point to God's will. They might say, "This was part of God's will for your life." God does not desire bad things to happen to people.

In Romans 11, Paul recognizes God's eternal love, asking the rhetorical question, "Has God rejected God's people?" Paul answers his own question in the strongest words he has, *may ghin'-oh-to: By no means!* Paul draws two possible conclusions about salvation and the irrevocable gift of grace. Either (a) God saves all Jews because some of them believe, or (b) God saves all people, including non-believing Jews.

What Is the Nature of Salvation?

Jesus dealt with his calling to be the Jewish messiah. Paul wrestled with the notion of Jewish people who rejected Jesus. In Matthew 15:21-28, a Canaanite woman asks Jesus to heal her daughter. At first Jesus resists her cries, saying that he was sent to the lost sheep of Israel. However, she persisted, wanting a scrap from the table of grace. So, he commended her faith and healed her daughter. His response pointed toward the later command to the disciples: that they were to become witnesses beyond Israel.

Shortly after the events of August 12, 2017, President Trump defended the white supremacists saying that there are "some very fine people on both sides."[5] In some ways, he is correct. In other ways, his comments are abhorrent and completely out of touch with the pain surrounding hate groups. First, how is he correct? Sin is universal (Romans 3:23). No one is all bad and no one is all good. Thus, the logical conclusion could be that there is goodness in everyone. If this is correct, then there would be "some very fine people on both sides." Second, his comments completely miss the point and the national mood.

Despite God's desire for harmony, we find ways to bring dissonance. In Charlottesville, Virginia, during the weekend of August 12, 2017, people fought with one another and created pain and destruction, and someone killed a young woman named Heather Heyer by driving a car into the crowd. James Alex Fields Jr. is accused of driving the car that killed Heyer and injured several others. Fields was part of the white supremacist rally.

Is there an alternative to hatred? Actions like Fields's car attack look like evidence of hatred. Listening is key to understanding. Daryl Davis exemplifies listening and engaging. He is a blues musician who has been befriending members of the KKK for more than thirty years.[6] As his friendship with them grows, he experiences people realizing that their hate might be misguided. After listening and establishing friendships, this one man pushes back against the forces of evil.

There are places where we should reserve judgment. Is this one of them? *may ghin'-oh-to*: *By no means!* We can be crystal clear. In any situation, when one side includes white supremacists, anti-Semites, or neo-Nazis, we can unequivocally say they are wrong and we should stand against them. We can apply this same response to terrorists—regardless of where they are from or their religion or ideology.

When a burglar breaks into a house, the inhabitants fear for their safety. Pointing out all of the good characteristics of the burglar as a person does not reduce the fear. Humanizing the burglar might help in the abstract. It might be a good approach for considering programs that help prevent people from entering a life of crime, but, in that moment, the people who live in the house just want to feel safe and do not want to feel violated.

Likewise, after a hate group disrupts life in an otherwise quiet, small city, sympathizing with the marauding band of white supremacists provides no healing. We can humanize them, but not sympathize with them. I cannot imagine the pain Heather Heyer's parents feel. Her parents are not alone. When we think about their pain, we should not sympathize with white supremacists. There is suffering around the world.

After the Rally

A few days after the white supremacists' rally in Charlottesville, NPR broadcasted a story about 16-year-old Brandon Martinez.[7] He was one of forty undocumented immigrants who were crammed into an unventilated trailer. Everyone in the trailer had paid smugglers to take them from Mexico to the United States. The smugglers and the driver of the truck abandoned them. Someone found them on a blistering hot day in a Walmart parking lot near San Antonio. Brandon was one of the lucky ones. He survived.

The news reported his father's struggle. His father works as a landscaper and is in the U.S. illegally. While he was visiting his son in the hospital, Immigration and Customs Enforcement tried to arrest him. The nurses and a lawyer protected him. I cannot imagine the horror his father experienced as he sat by his son's hospital bed.

Even though some people might have kept their focus on the white supremacists' rally, the world kept turning. Brandon Martinez fought for his life, and his story is one of 7.7 billion stories that continue each day. Around the world, people are starving for a scrap of grace. Paul says, "The gifts and the calling of God are irrevocable" (Romans 11:29). The Syrian civil war has claimed as many as 470,000 casualties. Just on the continent of Africa, there are currently nine ongoing conflicts that claim thousands of lives every year. These include the Somali civil war, the Libyan civil war, war in Darfur, and the Boko Haram insurgency.

How can we talk about God's love in the middle of all of this pain? How would we tell a child in Darfur that God loves her?

How would we explain God's irrevocable grace to parents who lost a child to Boko Haram? The problem of evil is related not only to human freedom, but also to the void of grace in this world. God loves everyone. God wants everyone to live in peace and to grow in faith. When people hurt one another, whether in Charlottesville or in Somalia, people do so despite God's continuous grace.

A Home That Has Never Known Sorrow

There is an old Chinese parable about a woman whose only son died. In her grief, she went to a sage and asked, "Can you bring my son back to life?" Instead of sending her away or reasoning with her, the sage said to her, "Fetch me a mustard seed from a home that has never known sorrow. We will use it to drive the sorrow out of your life."

She went off on her quest and sought a magical mustard seed. She found a splendid mansion, knocked at the door, and said, "I am looking for a home that has never known sorrow. Have you known any sorrow? This is very important to me." The person at the door said, "You've certainly come to the wrong place!" He began to describe all of the tragedy in their lives. Listening to the stories of their struggles and sorrows filled her heart with compassion. She tried to comfort the man as he told the story. He invited her in, and she found ways to help lift their burdens.

After staying for some time, she went on in search of a home that had never known sorrow. Everywhere she turned, she found one tale after another of sadness and misfortune. She became so involved in helping others cope with their sorrows that she eventually let go of her own. In her quest to find the mustard seed, her pain and suffering faded away.

One interpretation of the parable would be that God works through others and through service to others to comfort those who experience pain. Through this kind of experience, God's grace transcends the pain of death and destruction. It might not take the pain away, but we grow. Perhaps that does not seem like enough. Maybe we want an answer. But, like the Canaanite woman in Matthew 15, we must keep seeking, keep asking—and we can have faith that God will answer. And like the bereaved woman of the Chinese parable, the exercise of seeking can help us grow. It is not an intellectual exercise. When we put our faith into practice and move outside of our comfort zones, we grow and experience God's grace.

What Is a "Civilized World"?

One of the hallmarks of U.S. is the peaceful transition from one administration to another. Like the air we breathe or the landscape around us, taking U.S. democracy for granted is easy. Each presidential administration is unique, and, in some cases, the differences are striking. The Obama and Trump Administrations have radically dissimilar approaches. They look at the world in divergent ways.

We have addressed several phrases from Trump's inauguration speech. Overall, his speech revealed a clear worldview—one that is consistent with his campaign rhetoric and transactional worldview. Much of his rhetoric contributes to the pervasive polarization in the United States. Looking at how God can speak to the divisiveness means engaging with the polarizing language. It is messy work, but the effort can be a step toward overcoming differences and bringing people together.

Trump views the world as a zero-sum game. There are winners and losers. To be a winner means beating someone else. Instead of cooperation and mutual gain, Trump sees individual gain. In his inauguration speech, Trump said, "We will reinforce old alliances and form new ones—and unite the civilized world against radical Islamic terrorism, which we will eradicate completely from the face of the Earth."[8] What did he mean by "civilized world"? The countries of the North Atlantic? Europe, the United States, and Canada?

What about Australia and South Africa? Each of these areas have a majority white population. Does he include South Korea and Japan as civilized? What is the President's view of Bhutan? It is the only carbon-negative country in the world.

By saying the words "civilized world," we learn that he holds the antiquated view of some of the world as uncivilized. If uncivilized means Africa, a continent of fifty-four countries, the President does not understand the nature of civilization. The 1.1 billion people of the continent have produced twenty-two Nobel Laureates. Four of the ten fastest-growing economies in the world are on the continent. One out of every three people who live on the continent are middle class.[9] Sure, the continent has issues of poverty, inequality, and injustice, but so does the United States. We can relegate a phrase like "civilized world" to old books. When we read them, we can overlook them. These words are a relic of a Eurocentric or colonial worldview, like the sexist *mankind*, instead of *humanity*, or the racist *negro*, instead of *African American*.

Some of Trump's most ardent followers long for a golden heyday of the United States. They seem to long for something like the world portrayed on *Leave It to Beaver* (1957–1963) or *The Adventures of Ozzie and Harriet* (1952–1966). Both television shows portray an idealized America of the 1950s. Overcoming differences means acknowledging the sense of dislocation these followers feel. Yet as idealized as shows like *Leave It to Beaver* and *Ozzie and Harriet* seem, they miss the cultural richness of America, even within that era. Neither show portrays the depth of Appalachia and the Southern literary tradition. Both miss the growing Latino culture in the United States. Neither touches African American or Native American cultures, and, neither shows the diversity of American cities, like New York and Chicago.

Overcoming differences means being honest. Just as we must acknowledge the sense of dislocation or disenfranchisement felt by Trump followers, we must also be honest about the way different groups have treated one another. The white-only world of 1950s television is appealing for white males, especially those who do not have higher education. They can harken to an imagined past when they could graduate from high school, go to work, and make a living wage. To accomplish this lifestyle, people need living wages.

There is no civilized and uncivilized world anymore. We live in a postcolonial world. We live in a world of pluralities and differing viewpoints. We recognize various cultures for their richness and what people can learn from one another. Referring to the "civilized world" and longing for an idealized past might play well with white supremacists, but it means nothing to those of us who feel blessed to live in a diverse land. These phrases miss the truth, beauty, and goodness of seeing other nations succeed along with the United States. A truly "civilized world" would be one marked by inclusiveness, love, joy, peace, patience, and self-control.

Notes

1. Fred Rogers, *The World According to Mr. Rogers: Important Things to Remember* (New York: Hachette, 2014), 64.

2. Donald Trump, "The Inaugural Address," (2017), https://www.white house.gov/briefings-statements/the-inaugural-address/.

3. Harold S. Kushner, *When Bad Things Happen to Good People* (London: Pan Books, 1992).

4. Nicolas Berdyaev, *The Destiny of Man*, trans. Natalie Duddington (New York: Harper Torchbooks/The Cloister Library, 1960), 24.

5. Rosie Gray, "Trump Defends White-Nationalist Protesters: 'Some Very Fine People on Both Sides,'" *The Atlantic*, August 15, 2017.

6. Dwane Brown, "How One Man Convinced 200 Ku Klux Klan Members To Give Up Their Robes," *All Things Considered*, August 20, 2017.

7. John Burnett, "He Crossed the Border in a Packed, Unventilated Trailer and Survived," *Morning Edition*, August 17, 2017.

8. Trump, "The Inaugural Address."

9. Bhaskar Chakravorti, "It is time to get past the "single story" *about Africa*, Brookings Institution (2015).

5

New Beings in Christ

"So anyone who is in Christ is a new being…"
—2 Corinthians 5:17a

"The New Being is new in so far as it is the undistorted manifestation of essential being within and under the conditions of existence."[1]
—Paul Tillich

People tend to conform, to copy one another, and to gather in groups. Yet God calls each of us to be transformed into new beings in Christ. For Paul, worship and the Christian life are full-bodied experiences. We become "new beings," and worship is a tangible manifestation of faith in every part of our lives. We represent Christ when we are with our friends, at the store, at school or work, running errands, or at church. The faith journey impacts our relationships, the way we spend our time and money, and the way we think.

Spiritual growth means moving from the way we were to the way God wants us to be. This transformation is not a glum experience, but a vibrant transition into a rich, colorful life—one worth living. I can say, from personal experience, that this new life in Christ is one that enriches every single day. Let me be clear: becoming a new being is not some fallacious prosperity gospel. It does not impact our bottom line. It changes our approach to problems and our sense of self. The opposite of transformation into new beings is conforming. Conforming is what the world wants. It is not God's calling.

Consider the apostle Paul's life. What do we know? The New Testament is not a biography, but it includes clues about his background. For example, Acts 22:3 depicts Paul speaking. He says he was born in Tarsus as a Roman citizen. Born to Jewish parents, he was Jewish by birth and context. Paul's life began in conformity: he went to Jerusalem and studied the Torah under Gamaliel. This would be like following one's parents' footsteps and attending their alma mater. Paul's parents must have been proud. He was the son of a Pharisee (Acts 23:6). We have no way of knowing if he had a family connection with Gamaliel, but it seems that this would be like taking classes under your parents' favorite professor. He became a Pharisee. His parents, aunts, uncles, and cousins probably celebrated. His teacher was proud. His parents' neighbors were even proud. Today, they might say, "The kid done good."

What caused Paul to go off the rails? That is how his parents, neighbors, friends, and everyone else who celebrated his earlier achievements would have seen it. Everything looked so promising for him. Then, he had that accident on the road near Damascus. They might wonder why he could not just recover from the trauma and get back to being the Paul of whom they were so proud. Instead, Paul experienced transformation. He changed and became a new being.

Pressure to Conform

We all feel pressure to conform to the world around us and to meet certain expectations. For example, my friend Larry DiPaul, who was a Catholic priest, was saying his prayers one night when he felt God calling him to leave the priesthood. Up to that point, Larry's life had been one of both following God and conforming to his world's expectations. But, to leave the priesthood was a radical step that took him away from his life of conformity.

Larry took the step of following when God called him to leave the priesthood. He had some ups and downs. A couple years later, he was praying and felt God saying that there was a new calling ahead of him. The next day, he ran into a friend who said, "Larry, I've been trying to reach you. Would you be willing to be the new director of the Romero Center for Social Justice?"

This would be a hard job. The Romero Center is in Camden, New Jersey, in one of the roughest areas of the city. The job entailed spending time with the poorest of poor people in the country. Larry knew the task would often be thankless and difficult, if not impossible. He would work with addicted and homeless people, dealing with hopeless problems. He might encounter violence. Most importantly, it was the exact thing he knew God was calling him to do. Larry looked up at the sky and said, "See? This is why people don't pray anymore!"

If Larry had continued in his life of conformity, he would have missed the rich calling of working with people at the Romero Center. God's calling does not come when people expect it or are asking for it. Just as Larry's sense of calling did not come when he expected it, in Acts 9 we find the apostle Paul conforming to his calling as a Pharisee and was not expecting to encounter God. He was walking to Damascus because his job had led him to deal with a band of dissidents who followed an itinerant rabbi named Jesus. The blinding light on the Damascus road gave him a choice: continue conforming or be transformed. Paul said yes to the Risen Christ. He stopped conforming to this life and followed God on a new path. Imagine how his parents and everyone else felt about this new calling.

Our culture inundates us with messages telling us to conform to the world, but in Romans 12:2, Paul says, "Do not be conformed to this age." A different Paul—the theologian Paul Tillich—writes, "This warning of Paul is significant for all periods in history. It is urgently needed in our period. It applies to each of us, to our civilization, to humanity as a whole."[2] Tillich wrote this in 1963, and it is no less true today. In our age, marketers want us to conform to trends, to have the latest clothing, phones, cars, and other material possessions. I heard a radio advertisement for donating one's car for public radio. The pitch was, "Stop embarrassing your kids when you pick them up from school with that old car. Donate yours and buy a new one."[3]

Our act of conforming to this age extends into every aspect of our lives. Relationships become disposable and community ceases to be interpersonal, often limited to a virtual reality. Virtual reality is a contradiction in terms. *Reality* is "the world or the state of things as they actually exist," and *virtual* is "almost or nearly as described." The apostle Paul says, "Do not be conformed to this age." Yet another Paul, Paul Achtemeier, writes about Romans 12:

> It is clear from the opening verse that grace is to affect the whole of human life. In language, reminiscent of 6:12-13…, Paul tells his readers that their proper response to their Creator is the shaping of their total lives by [God's] gracious will. Like the burnt offering given [entirely] to God, the Christian is to be a total sacrifice to God, and that sacrifice is to consist of the whole of life. That, says Paul, is the logical response… to the history of God's grace…[4]

Instead of shaping our lives to the world around us, Paul Achtemeier points out how Christians should be shaped by God's "gracious will" and offer the self to God.

"Do not be conformed." These four, simple words challenge our entire civilization. We depend on conformity. It gives us a sense of identity. People can be conformed not only to a group, but also to themselves. A fitness buff can become so accustomed to working out that she loses her freedom and becomes a conformist to fitness. Social media breeds conformity with "likes" and "shares." Video games draw people into artificial worlds. These are kinds of slavery to self, and, it can happen in any hobby, interest, vocation, or relationship.[5] Yet, God wants transformation.

Tillich refers to transformation in Christ as the "undistorted manifestation"[6] of reality. When we separate from other people and form our tribes, we fall short of God's transformation. God calls humanity to unity. Instead, we find ways to separate ourselves from one another. One of the most insidious separations in society today has become our radically partisan political parties. In the U.S. today, it seems that Democrats and Republicans cannot work together. The resultant void of accomplishments plagues Congress.

In this dynamic, compromise becomes a dirty word. The way of the world politicizes every issue. Instead of carefully thinking through the pros and cons and weighing the possible outcomes against one's ideology, identifying each position with a party is far simpler. For example:

> Tighter border security = Republican
> Deferred Action for Childhood Arrivals
>   (DACA) = Democrat
> Fewer regulations for business = Republican
> Better environmental enforcement = Democrat
> Pro-life = Republican
> Pro-choice = Democrat
> Second Amendment Rights = Republican
> Gun safety (or control) = Democrat

This list could go on. Yet, few people are as monolithic as this dichotomy in the U.S. represents. The apostles Peter and Paul could barely stand to be in the same city. Galatians 2:11 says, "When Peter came to Antioch, I opposed him to his face." Yet, in Acts 15, they seemed to have resolved their differences and found common ground in Christ. As new beings, they either reconciled their differences, or acknowledged them and continued their work. Christians today can follow their example. To not conform to this world means standing against humanly created divisions.

Alternative to Conforming to the World

What is Paul's problem with conforming? When we conform to this world, we put something above God. Paul wants Christians to give their entire self to God. The faith journey involves our entire self. Achtemeier writes, "Life under the lordship of God means a life under the structuring power of Grace overcomes slavery to self, liberating us for a life of faith."[7] Grace structures life and systematizes our response to what we see. When the world pulls people into tribes, grace confronts conforming to the division.

My friend Larry DiPaul showed me how to confront conforming to divisions. He was the former priest who became the director of the Romero Center in Camden, New Jersey. A few years ago, he was diagnosed with cancer. After struggling for less than a year, he died. If he had not listened to God and been open to transformation, he would have missed the incredible joy he found at the Romero Center. The apostle Paul echoes Larry's joy in faith. From the depth of prison, he and Silas sang hymns. These were not songs of regret. They were songs of praise to God. They were not songs of conforming to the world, but songs of transformation.

We can find another approach to living as a new being in the Hebrew prophetic tradition. Hosea, the first of the twelve Minor Prophets in the Bible, expresses the theme of God's love for ancient Israel. In fact, God loved Israel with the same kind of love marriage partners exhibit toward one another. This theme is emphasized through the ongoing metaphor of Hosea's own marriage.

A 1929 play called *The Marriage of Hosea: A Passion Play in Three Acts* is beneficial for understanding the themes of Hosea. By describing it as a *passion* play, the pseudonymous twentieth-century author Izachak[8] understands the ancient prophetic book of Hosea very well. The word *passion* is a multivalent word; it has many values, meanings, or appeals. The characters in the play all have fiery relationships. Hosea's wife burns with lust as she chases her lovers. Her husband burns anger as he lashes out at his wife to rebuke her and bring her back. Passion embodies Hosea's suffering when tortured by her, and passion grips the love between the couple as they reconcile and commit to each other again.

What we find in the book of Hosea is a glimpse into the deepest emotions of life. This is not a tepid book. It is not a distant theology. It requires emotional engagement. When we recognize the passion of Hosea and the connection to strong feelings, the ancient text can speak to the political divisions today. People confront an issue from different sides. As the confrontation heats up, passion lurks beneath the surface. Why would people violently lash out if they did not care? They do care. They care about issues. They care about their understanding of right and wrong.

This passion—Hosea's deep feeling—can only be appreciated when we step inside this tempest of ups and downs, and not just slight ups and downs, but those great swings of absolute elation to the very pit of despair. In order to get something out of Hosea, we have to prepare ourselves. A light reading will just not do!

The Need to Engage

Hosea demands engagement. I like Hosea because it speaks to twenty-first-century divisions. Hosea is not Sunday-after-noon-sitting-in-a-rocking-chair-on-the-porch theology. It is full of the realities of the world, like a perverse priesthood (4:1-6), the politics of self-destruction (5:8–8:14), and the people's history of infidelity (9:1–11:11). The world, the headlines, the hurts, the crimes, the weddings, the funerals, the hospital-emergency-room visits—these things do not exist in a cold, emotionless world.

Behind every headline is someone with deep feelings. A business failure equates to a person's lost dreams. A corrupt politician facing trial feels the pain and desperation of having let everyone down. The same is true for both victims and perpetrators of crime, for brides and grooms in weddings, for mourners at a funeral. All of these people experience the true depths of human emotion and the great feeling of daily life. And, the question in all of these situations is: how do we respond? Do we follow our gut instinct? Do we follow the crowd or what other people think we should do? Or, do we acknowledge the deep love God has for us and follow our Lord? Hosea 11 is a record of one such emotion-ridden interaction.

Hosea 11:1 begins with a child who is deeply loved: "When Israel was a child, then I gave my love, and called my child out of Egypt." The child is called, and the more the child is called, the further the child drifts away. The child is reminded of the deep love of the parent. "They called to them, so they went from them" (Hosea 11:2). The child is reminded of being gathered up in the parent's arms (Hosea 11:3). The child learns that the parent was the one who provided healing when the child needed it. The child hears again about being lifted like an infant to the mother's cheek and being fed like a baby.

Then, Hosea 11:6 turns to anger. "The sword will fall on their cities and will destroy the bars of their gates, and will put an end to their plans." We have a display of the *passion* mentioned above. "That's it! Get out! I've had it! I've called and called, and if you are so bent on turning away from me, then fine. Go!" However, the passage doesn't end with this burning anger. Again, it sounds like a mother's voice (11:8), "How can I give you up, my child? How can I hand you over? How can I treat you badly?" The speaker has calmed and remembers the deep love expressed earlier. "I will not execute my fierce anger. I will not destroy you" (11:9).

This passage contains one of those great images we find of the biblical mothering God. This image gives us a fuller picture of the relationship God desires to have with humanity. The central theme in Hosea is God's love for Israel. To apply this book to our lives and the situation in twenty-first-century United States, we could make the text personal—the central theme is God's love for us. The maternal images, just like paternal images for God, are metaphors, useful illustrations to help us better understand God. We could say that all images of, or words about, God are useful ways to help us better understand God and the world.

Suddenly the prophet's voice changes. No longer is it a mother speaking to a child. We, the listeners, are reminded: this is God! "They shall go after the Lord…they shall come trembling like birds from Egypt, and like doves from the land of Assyria; and I will return them to their homes, says the Lord" (11:10-11).

God is like a mother calling her child from the front porch. Then, the child willfully runs away, as if looking up and seeing the mother calling, yet deliberately ignoring the calls. The loving call changes to frustration. The frustration is real. It is like that feeling most parents have had: *I don't know what to do with you!* This is not a glossy image of some distant deity who cannot understand human emotion and experience. God experiences our pain and our reality. This is a God who knows us and who wants to know us, one who continually seeks to deepen a relationship with us.

A Return to Egypt?

When the nation misbehaved, prophets like Hosea threatened a return to Egypt. How does history function for present life? How does remembering the past impact the present? When the white supremacists gathered in Charlottesville in 2017, some rallying cries focused on tradition. Someone could argue, "General Lee's battle flag is not a symbol of hate. It symbolizes *my* heritage." Whenever life grew unbearable in the wilderness, the ancient Israelites longed for their days of slavery in Egypt. They remembered having food and shelter in Egypt—even though they had to work as slaves.

For those who see the racism of the Civil War era, mentioning Lee's battle flag harkens a dark time in U.S. history. For those who feel dislocated by a changing world, Lee's battle flag may symbolize a better time—even if that better time is imaginary. It is an idealized past. Racism and slavery might not be an explicit wish for every person who flies the battle flag, but the history that accompanies the flag is not silent or impartial. The flag carries weight. It represents a story. Flying it means accepting the weight of that story. Honest conversations uncover this meaning.

Why does Hosea speak of the people's return to Egypt? It is a threat, like seeing a parent continue to call from the porch, "Come on, come in." Why does the child ignore the parent? As a bystander, it is hard to fathom. Contemporary readers of Hosea hear calls to truth, beauty, and goodness, yet at some level, everyone turns away. We do it in different ways. We know we should eat healthy but do not always do it. We know we should drink more water but do not always do it. When considering Lee's battle flag and everything it symbolizes, I wonder why someone would raise a symbol that signifies hatred, racism, and slavery to so many people? Is there no other way to celebrate or honor one's heritage? Do they not see how it hurts other people? The symbol is divisive and does not represent any kind of universal love.

When the evidence is clear, the bystander expects rational behavior. People know that drinking more water is healthier, so we are not surprised when people carry water bottles. For the child, returning to the parent makes sense. When we apply this same logic to symbols, on the one hand, it seems like the person who flies

Lee's battle flag should lower it. On the other hand, when we apply this logic to behavior, all Christians should be transformed into new beings in Christ. Yet, transformation does not happen often or consistently. Christians still sin, children disobey parents, and people thoughtlessly raise symbols invoking the ire of their neighbors.

For the ancient Israelites, the idea of a return to Egypt means an actual uprooting from the land. It is the threat of captivity in a foreign and hostile country (Hosea 9:3; 11:5). This kind of threat includes actual slavery in exile. It means overseers using whips on the backs of the re-enslaved people and placing them in squalid living and working conditions (cf. 10:11). Childhood vanishes in slavery (cf. 9:12). Freedom, dignity, and humanity depart with the arrival of slavery. Hope, despair, and prayers for another Moses to lead an exodus come back. Stated plainly, no one wants this.

Throughout chapters 9–11, Hosea summons the people to "remember" their history. In Hosea 11:1, there is the reminder that God freed them from slavery in Egypt. Both Hosea 9:10 and 11:3-4 reminds the people that God chose them and protected them in the wilderness. Remembering history can be both a source of gratitude for the good and a reminder to avoid repeating mistakes.

When we read in this way, Hosea speaks to the politicized world of the twenty-first century. Learn from history. Take religious and political decisions seriously. As contemporary voices seek to find scapegoats for various problems, study twentieth-century history. During the Holocaust, Nazis killed six million Jewish people. They started by blaming Jews for every problem. Blaming any group, such as immigrants, Muslims, Mexicans, Russians, or any other boogeyman, dehumanizes the individuality of each person and misses the complexity of actually addressing the problem. Racism, intolerance, and various forms of ethnic cleansing continue to confront humanity. Hosea provides a warning. If we turn away from problems, if Christians ignore the transformative power of Christ, we can also "return to Egypt" in our own way. As new beings in

Christ, people can look beyond the metaphorical Egypt. Instead of looking back to the past, Christians can lean into the notion of transformation to new beings and look to the future.

Notes

1. Paul Tillich, *Systematic Theology, Volume 2: Existence and the Christ* (Chicago: The University of Chicago Press, 1957), 119.

2. Paul Tillich, *The Eternal Now: Sermons* (London: SCM Press, 1963), 115.

3. Catherine Fenollosa, "Turn Your Car into the Public Radio Programs You Love," *NPR Extra*, November 14, 2013.

4. Paul J. Achtemeier, *Romans*, ed. James L. Mays, Patrick D. Miller, Jr., and Paul J. Achtemeier, Interpretation (Louisville: John Knox Press, 1985), 195.

5. Tillich, *The Eternal Now: Sermons*, 116.

6. Tillich, Systematic Theology, Volume 2, 119.

7. Achtemeier, *Romans*, 195.

8. Izachak, *The Marriage of Hosea: A Passion Play in Three Acts* (New York: Halcyon, 1929).

6
Trust as a Reconciling Act

"You never really understand a person until you consider things from his point of view...until you climb inside of his skin and walk around in it."[1]
—Atticus Finch

"The greater the contrast, the less perfect the actual reconciliation, so that when all is said and done there is often no reconciliation but rather enmity."[2]
—Søren Kierkegaard

"Some trust in chariots, and some in horses, but we trust in the name of the LORD our God."
—Psalm 20:7

In *To Kill a Mockingbird*, Mayella Ewell had no friends and felt lonely. The Great Depression chased away not just material wealth but hope too. For some people, it left a poverty of the soul. During the long, hot summer days in Maycomb, Alabama, Mayella looked for a friend in Tom Robinson. The problem was that in the 1930s, in Maycomb, Alabama, Tom and Mayella could not be friends, or anything else. Mayella was white and Tom was African American. Mayella did not see it as a problem and sought Tom's affection. Tom knew this could go only one way. He rejected her advances, and in the words of William Congrove,

> Heav'n has no Rage, like Love to Hatred turn'd,
> Nor Hell a Fury, like a Woman scorn'd.[3]

Mayella's loneliness broke through racial barriers. Then, her pride pushed her and Tom worlds apart. Mayella accused Tom of attacking her.

Atticus Finch was the attorney who tried to defend Tom. Tom's voice rings out in the opening words of the psalm, "Vindicate me, O Lord, for I have walked in my integrity, and I have trusted in the Lord without wavering. Prove me, O Lord, and try me; test my heart and mind" (26:1-2).

Who stands with the boldness, even arrogance, to say, "Vindicate me, O Lord"? One who is innocent. When accused, innocent people demand a trial. Guilty parties might seek a trial and a legal loophole, but only innocent people say, "Vindicate me, O Lord!" I do not know if Harper Lee thought about Psalm 26 when she wrote *To Kill a Mockingbird*. And in Psalm 26, the root cause for vindication is unclear. We do not know if this psalm is about a trial or false accusation.[4]

According to Paul Mosca, this is the private prayer of a priest preparing for worship.[5] It sounds like a plea: "I want to serve you, God. You know me. You know what I do, how I act. Please find me worthy of your presence."

Mosca's approach works for people seeking reconciliation. If I approach a friend who holds differing ideas, my first words might be, "You know me. You know what I do, how I act...Please find me worth your engagement. Please find value in our reconciliation."

Too often, readers of Psalm 26 connect it with the Pharisee in Luke 18:11. The Pharisee says, "God, I thank you that I am not like other people." Jesus is critical of his prayer. The Bible lacks stage direction or notes about tone, but the Pharisee could have said *other people* with a certain implication in his voice. Connecting Psalm 26 with self-righteousness suggests something that is not necessarily there. Yet, there are two troubling aspects of Psalm 26. First, it seems self-righteous. Second, separating the wicked from the righteous sounds elitist. All of this hearkens to the Pharisee in Luke 18, and Jesus is critical of him. We do not want to be like the Pharisee. However, this kind of reading ignores the possible context of accusing someone who is innocent. The jury in Tom Robinson's trial ignored the facts and assumed Mayella told the truth. Tom sought vindication in the truest sense of the word. Mayella sought something closer to validation or vindication in the loosest sense of the word. Although "with God all things are possible," in the novel there is no context for or possibility of reconciliation.

What is the truth? Where do we find it? How trusting can we be? In the following pages, we will explore the nature of trust and its relationship with truth. We will look at context, the need to stay together, and how we are worth more together than apart.

Context Matters

Context matters. Søren Kierkegaard brings out the challenge of reconciling and the importance of context. Context plays an essential role in understanding two contrasting positions. Actual reconciliation becomes more difficult with a greater difference between two positions. For the psalmist, the context is innocence. For Tom Robinson, the context is wrongful accusation. For people in a polarized political world, deep-set ideology frames the backdrop of reconciliation.

The psalmist's plea, "Vindicate me, O Lord," suggests a deep trust in God. In other words, *Lord, no matter what is happening, I trust you.* The psalmist knows that God is greater than various problems or circumstances. Maybe everybody is not in Tom Robinson's shoes. Not everyone has someone falsely accusing them. But this plea is a basis for trusting God. The psalmist believes that God is interacting with humanity. We are not alone. God cares, and God acts in people's lives. When we pray, our words do not hit the ceiling and bounce back. God is with us.

The psalmist suggests a divine ethic: "I do not sit with the worthless…I hate the company of evildoers." There is a good and a bad. How people behave matters. James Mays writes: "Psalm 26 reminds us, then, that there is a legitimate form of separatism. Not anything goes! God opposes evil. Those who submit their lives to God's sovereignty will be different from those who follow only the direction of the self."[6] The faith journey has a form. It has direction.

What is this direction? Is it, as the prophet Micah 6:8 says, to "do justice, seek mercy, and walk humbly with your God"? In Matthew 16:24, Jesus says, "If any want to become my followers, let them deny themselves and take up their cross and follow me." To live out the psalm means having a tangible faith. It means that people could recognize God in us. The way we speak, the way we act, and the way we live our lives should reflect the plea in this psalm. It should reflect our trust in God.

How do we trust God? Not, *do we trust God?* The latter is a simple yes or no question. We can say, "Yes, I trust God," and be done with it. But to ask, "How do we trust God?" implies something deeper for our lives. What do we do to provide evidence for our trust in God? Do we lay ourselves bare, saying, "Search the depths of my soul, Lord, because we know you already know, and I feel confident that my life reflects your love"? Then, as we bare our souls before God, we can live out our trust in tangible ways.

In *To Kill a Mockingbird*, Atticus Finch did not convince the jury to exonerate Tom Robinson. But God is bigger than the fictional lawyer in Harper Lee's book. Christians who seek to reconcile with others can trust and know that God is God and capable of addressing injustices.

Remain Together

Trusting means remaining in contact with one another. We cannot trust those with whom we have no contact. The Bible is a message to a particular people at a particular time and place. In each story, the people interact. Matthew 18:15-20 addresses conflict.

In the larger section around Matthew 18, Jesus deals with life in the ekklesia, which we translate as church. Ekklesia was a common noun in the first century. It is a compound noun, like toothpaste or household. In some cases, the two words describe the meaning of the new one. Toothpaste is paste to use on your teeth. In other cases, the two words do not describe the meaning of the new one. A household is not a house that we hold, but a house and its occupants. Ekklesia is *ek* (out of) and *kaleo* (I call), but it does not mean "called out." It means "a popular meeting, especially religious" or it could be translated as "assembly." Elsewhere (Acts 19), ekklesia is translated "assembly."

We read the word "church" and picture a building with a steeple, and when we open the doors we find the people. But Jesus had an entirely different image in mind. At the beginning of Matthew 18, he says that people whom the world sees as important or significant should be humble like children. This is a powerful message for weak and powerless people, but it might feel confrontational to those who have power or a high position. Then, he addresses his disciples.

Disciple means "learner," and we all want to be Jesus' learners or followers. Jesus has great expectations of his followers. Saying yes to Jesus means accepting the whole package. This includes the positive lessons about God's love and grace. It also includes the difficult lessons about humility and putting God above the self. Then, he cautions would-be disciples against doing anything to make someone else stumble. Each one of us is responsible before God for our behavior.

In twenty-first-century America, people crave individualism. Unlike the people of the gospel, many of us are self-reliant today and live in a me-first world. When someone is unhappy at church, for example, they leave and find another church. People in the ancient world often lived communally, sharing everything. When they disagreed with one another, they could not just leave. When early Christ followers were upset with one another, they could not find another church. They could argue and cause trouble, but there was not another church down the street. To become Christ followers, many had given up the religion of their birth. This decision fractured family and community ties. Now, they had a new one.

We are all at different places in our spiritual journeys. Because of the freedom God gives us, we arrive at different conclusions. Add geographical background, upbringing, experiences, and education to the mix, and we are a bunch of distinctly individual people. In our uniqueness, we think differently from each other. Despite these differences, Jesus calls us to unity, to live in faith together.

Unity is especially difficult in an age that values individuality. Individuality breeds varying positions. Because of the value we place on individuality, those unique positions quickly become unwavering. Two opposing, unwavering positions create conflict. When people become entrenched in their own positions and do not consider the other person, oppositional positions devolve into win/lose. Unwavering positions become stumbling blocks for those who, through a lifetime of different experiences, have landed at a different position and disagree.

Jesus feels so strongly about these stumbling blocks that he says, "It would be better to have a millstone around your neck and to be thrown in the sea" than to cause someone to stumble (Matthew 18:6). He addresses morality in stark hyperbole. "If your foot causes you to stumble...cast it off...for it is better to be maimed than to face eternal fire" (Matthew 18:8). Discipleship is hard. And that is Jesus' point. His calling is about transformation. We set aside our selfish ways and turn ourselves completely over to God.

What does "stumble" mean in this context? Falling down during a faith journey. It can also mean getting distracted and losing focus on growing with God. How does disagreeing with others cause someone to stumble? The answer is the way we disagree. Different people approach disagreements differently. Keeping the disagreement in context and remaining engaged can limit the likelihood of becoming a stumbling block for someone else.

Communication is a key element in trust building and avoiding becoming a stumbling block. Clear communication is honest and "speaking truth in love" (Ephesians 4:15). Communication means spending time with other people. It means listening as well as speaking. It means hearing the words other people say, considering them, and trying to understand where the other person is coming from. In this way, communication fosters empathy and erodes the sclerosis bred by distrust.

Worth More Together

We are worth more together than we are separately. But, when we come together, we bring our culture, in addition to individuality, which is another characteristic that distinguishes our culture from Matthew's audience. In the twenty-first century, even though postmodernism pervades our thought, many people remain children of the Enlightenment. Reason and rationalism dominate worldviews. As I argued above, Matthew's audience valued community and interdependence. This is why Paul uses his communal worldview in 1 Corinthians 12:12-26. The church is a body. Individuality provides different parts to the body—not disparate pieces that barely fit together, but valuable parts of one whole. Each part maintains its uniqueness, like a pot of stew. Meat, potatoes, carrots, and onions retain their shape and individual flavor, but they come together to make a richer flavor than the individual parts. To push the metaphor further, each ingredient gives its best to the mixture.

The church is full of people. Everyone can give their best to the mix of the church. But the church is not a stew. The people inside relate to one another. On a spiritual level, they relate like the most intimate relationship between two spouses. This is why the people in the church are the bride of Christ. We are God's ekklesia (religious assembly) and God's hands and feet in the world. We must trust one another, come together, and work for reconciliation. Yet, we come together as our sinful selves. Each one of us stands in need of God's grace and forgiveness. In addition to God's forgiveness, there are times when we need forgiveness from one another.

Matthew 18:15-20 is about reconciling when we fail at living as God's ekklesia or church. This might be one of the most common sins of the church today. Instead of following Jesus' example for reconciliation, we leave the church (either quietly or otherwise) and find another one. We leave one tribe to find another one we think fits more closely with how we see the world. There are many reasons for leaving rather than reconciling: pride, ego, misunderstanding, ignorance of a problem, or doctrinal differences. Church grievances are not usually interpersonal—they are theological because they hurt the body of Christ.

Earlier in Matthew (5:23-24), Jesus says, "When you are offering your gift at the altar, if you remember that your brother or sister has something against you, leave your gift there before the altar and go; first be reconciled to your brother or sister, and then come and offer your gift." The idea is one of making peace, living in harmony, and remembering what it means to be God's people. It is too easy to find things to criticize. It is too easy to find people to dislike.

Notes

1. Harper Lee, *To Kill a Mockingbird* (New York: J. B. Lippincott & Co., 1960), 39.

2. Søren Kierkegaard, from"The Morning Bride," *The Concept of Irony*, trans. Howard V. Hong and Edna H. Hong, vol. 2, Kierkegaard's Writings, (Princeton: Princeton University Press, 1989), 297.

3. William Congreve, *The Works of William Congreve*, ed. D. F. McKenzie, vol. 2 (Oxford: Oxford University Press, 2011), 60.

4. This particular psalm is elusive. Commentators disagree about its purpose and connection with other psalms. Some say it is connected with Psalms 7 and 17, in which the psalmist seeks justice. Others argue that it is similar to Psalms 15 and 24, and it is an entrance liturgy. Still others find a connection with psalms petitioning God's help against the wicked. See James L. Mays, *Psalms*, ed. James L. Mays, Patrick D. Miller, Jr., and Paul J. Actemeier, Interpretation (Louisville: John Knox, 1994), 781.

5. Paul G. Mosca, "Psalm 26: Poetic Structure and the Form-Critical Task," *The Catholic Biblical Quarterly* 47, no. 2 (1985).

6. Mays, *Psalms*, 783.

7
Go with God

"Theologians have suggested that prayers of supplication should be replaced by thanksgiving in order to avoid magic connotations. But actual religious life reacts violently against such a demand. People continue to use the power of their god by asking god's favors. They demand a concrete god, a god with whom people can deal."[1]

—Paul Tillich

"The LORD said to Moses, 'I will do this thing that you have spoken; for you have found favor in my sight, and I know you by name.'"

—Exodus 33:17

"Moses stretched out his hand over the sea, and the LORD caused the sea to go back by a strong east wind all night, and made the sea dry land, and the waters were divided. The children of Israel went into the midst of the sea on the dry ground, and the waters were a wall to them on their right hand, and on their left."

—Exodus 14:21-22

Once upon a time, a man named Anen watched the sunrise. As he watched, he looked down at his sandals. They were well-worn. His mind drifted back to the day when another soldier brought new sandals to him and some other men. He was excited, but they were stiff and took some time to break in. He thought philosophically about how life is a bit like those sandals and it takes time to understand things.

On this particular day, the sun came up over Anen's enemy. In the distance, he could see their camp by the sea near Baal-zephon. This was an enemy he did not know. As a soldier, he followed his orders. They had pursued this fleeing people, and now, his enemy had their back to the sea. The battle would soon be over.

Anen felt hopeful. His spear weighed heavily in his hand. Then, someone relieved him from standing watch, and he ate his breakfast. It was the same thing every day—bread. But, in ancient Egypt, he had never heard of peanut butter or jelly, and Anen was happy that the bread was not too stale. Anen's enemy, the ancient Israelites, began their journey to freedom (Exodus 12). Most people know the story of Exodus from the perspective of the ancient Israelites. As a lowly soldier, Anen did not know about his enemy, why they fled, or what the battle was about. He followed orders.

God Acts in the World

Sometimes, we understand following God. Other times, we must put ourselves in someone else's shoes to understand how to follow God. When we do, we can better understand what going with God means. Paul Tillich writes about the idea of asking God to do something. Instead of following God, people confront the temptation to direct God. That is, we pray, "God please do this…protect her…heal him…deliver me…" God is beyond human ability to manipulate.

Each of us is on a path. We are on a journey. On our path, we can see the world around us. Friedrich Schleiermacher writes, "The world is a work of which you survey only a part, and if this part were perfectly ordered and complete within itself, you would not be able to formulate any lofty concept of the whole."[2] Anen, the fictional character in the opening of this chapter, like us, can see his world. But, he can see only a part of the big picture. Likewise his enemy, the ancient Israelites, could see only their world. What kind of compassion would have filled their hearts if they could see the big picture? How would Anen feel if he knew them? What kind of compassion would fill our hearts if we could see a bigger picture? How would gaining perspective help overcome politicization in the twenty-first-century United States?

The exodus story is about God at work in the world. We are, in some ways, like the Israelite sojourners. We do not know our pursuers and are limited in knowing about the big picture. Despite news inundating daily life in the twenty-first century, we can be frighteningly ignorant about what makes other people tick. More and more, people form like-minded circles. Within those circles, we assume that we agree with one another. Gradually, we lose the ability to hear varying viewpoints and empathize with another person.

Exodus 14 is not about Egyptian or Israelite history. It is about God seeing and acting in the world. Jon Sobrino writes, "God never appears as a God-in-Godself, but as a God for history, and, therefore, as the God-of-a-people."[3] God's promise, "I will be your God and you shall be my people" (Exodus 6:7), is relational. God's revelation is always in relation to people. We do not experience God in isolation. The exodus was not human-focused, human-inspired, or human-accomplished. God heard the cry of the people and responded. "God is a God-*of*, a God-*for*, a God-*in*, *never a God-in-Godself*."[4]

Exodus 14 does not begin with history or easily verifiable details. It begins with something more powerful: the Lord speaks to Moses. This is a God-*of*, a God-*for*, and a God-*in* history. The chapter continues retelling about God transforming Pharaoh's perspective. In Exodus 12, Pharaoh was desperate for the troublemaker

Israelites to leave. Now, he has a radical change of heart. The scene shifts to Moses and the Israelites. They see Pharaoh's army coming. They see soldiers like Anen and are terrified. "Did you bring us here because there are no graves in Egypt?!" (Exodus 14:11). The people who complain mention the name Egypt five times. Walter Brueggemann writes, "It is the only name they know, the name upon which they rely, the name they love to sound. In the speech of the protesting, distrusting people, the name of [the Lord], however, is completely absent."5 Fear has driven them away from their faith. They do not understand a God-*of*, -*for*, and -*in* the world. They want something familiar, like a god that resembles their own worldview.

Moses tries to assuage their concerns: "Do not fear," he says. "Stand firm, and see what God is going to do." Moses stretches out his hand and the Lord drives the sea back. This is an example of Moses standing firm in God's work. We suspend explanatory suspicion and have faith that this is a story about God at work in the world and relating to humanity. God's people walk across on dry land. When Pharaoh's army pursues them, Moses stretches out his hand and the seas return, crushing and drowning Anen and his fellow soldiers.

God Delivers People Together

Many of us have heard since childhood about God delivering the Israelites. For many people in twentieth-century United States, Charlton Heston epitomizes Moses leading the children of Israel across the Red Sea in the Exodus. I have always thought of the story from the side of the Hebrew people, but what about the Egyptians? What does the story look like from their perspective? My heart breaks for them. Maybe I missed the point. Or, maybe not. Where we stand matters. Who we stand with matters. We can either go with God, or we can go our own way.

In Charlottesville, Virginia, on August 12, 2017, people in the city had to make a choice. They could stay home (most did), rally with the white supremacists, stand against the white supremacists, or show up but stand aside. This last category is difficult to pin down. It is where I stood. Is showing up and standing opposite the white supremacists a de facto stand against them? Some of my clergy friends gathered in a church to provide support to the counter protestors. Is this yet another category?

What about those who did not know where to stand? Again, I put myself in this category. Confusion reigns at these types of events. I know that I stand against the KKK, white supremacy, and neo-Nazis. But, how do we confront evil directly? Sun Tzu would scoff at this kind of direct confrontation with power. Instead, *The Art of War* directs people to know themselves and know their enemy. It says to attack where the enemy is weakest, yet "The supreme art of war is to subdue the enemy without fighting."[6] Thus, is it wise to directly confront people who have prepared for a fight? No. Doing so accomplishes nothing.

In Exodus, it is God, rather than people, who pushes back the water. God lets it go again. One of the main points in Exodus is our witnessing to a God-*of*, -*for*, and -*in* the world. The Israelites fail to recognize who God is. In their moment of pursuit, the Egyptians confess, "The Lord is fighting for them and against us."

We cannot know the hearts of the ancient Egyptians. However, seeing the story from their perspective can foster empathy in us. We cannot know the true motivation of another person. We cannot know the hearts of the KKK, white supremacists, and neo-Nazis who protested in Charlottesville. Therefore, we cannot judge the people who protested. We can criticize their actions but not them as God's creation.

Instead of focusing on the other, we can ask questions of ourselves. Are we going with God? Are we making decisions to move as God directs? Do we let faith dominate our lives? The fictional

Egyptian soldier Anen might never have heard of Yahweh. He might not have known about going through life with a God-*of* him, a God-*for* him, and a God-*engaged-with* him. Do we? How do our lives reflect our knowledge of God?

What Is Fair?

I have a friend named Jerry who owns a business. It is not a large business, but it is successful enough to support him and his family. Jerry's business is large enough for him to have several employees. He gets to choose who he hires. He decides how much he pays his employees. Once, one of his employees looked through the files when Jerry was out of the office and learned how much her coworkers earned. She was incensed when she discovered he paid a particular coworker more than her. She confronted Jerry, demanding a raise in pay: "Either pay me more, or I quit."

He said no to the request for a raise.

Did Jerry do what is "right"? Or, did the employee have a just claim on greater wages? There is more information to this story. The coworker who earned more possessed a specific and much-needed skill set. The one demanding a raise did not. Also, was she unjustly paid? The answer is no. My friend paid a fair wage for her skill level and work.

In the Gospels, Jesus used stories to explain God's kingdom. In Matthew, Jesus says, "For the kingdom of heaven is like a landowner who went out early in the morning to hire laborers for his vineyard." His listeners feel right at home. Then, Jesus adds some unfamiliar details, like the landowner went to the market. The listeners might ask, "Where is the manager? Why did the landowner go?"

But the landowner did go to the market and entered into a specific agreement with the first laborers. The landowner pays one denarius, which is a subsistence wage—not to imply that the

landowner was taking advantage of these laborers. It was probably a fair wage for unskilled laborers, but one denarius would not
be enough to support a family. Remember this agreement. It is crucial in the final scene of the parable.

The next detail that would have made Jesus' audience say, "Wait
a minute," was the landowner's repeated trips to the marketplace.
Matthew does not tell us why no one hired those who were "standing idle" earlier. The landowner hires the first group of workers
based on an oral contract for the accepted amount—one denarius.
Now, the landowner promises the next group what is "right." This
word, dik'-ah-yos, means equitable in character or act, and by
implication, it means what is holy or righteous. (This is the same
word used in Matthew 1:19 to describe Joseph when it says,
"Joseph, being a righteous (dik'-ah-yos) man and unwilling to
expose her to public disgrace, planned to dismiss her quietly.")
That is the amount the landowner will pay when it says the next
group will get what is "right." The first group has a verbal contract, but these new arrivals can only trust in what the master sees
as just or right. In fact, both groups depend on the trustworthiness
of the landowner.

In Matthew, the landowner returns to the marketplace and keeps
hiring laborers. These questions about the story remain. Why did
the landowner go? Why not send a manager? Why keep going
back? Why not hire enough at the front end? Why were these people still available to work at the end of the day? The answers to
these specific questions are not as important as seeing the way the
landowner treated the workers with righteous equity. The
landowner went out of care for the workers. A manager, in this
story, would not symbolize the deep concern the landowner had
for the workers. The landowner kept going back for the same reason as going in the first place: care for the workers.

In a contemporary context, a poor Nicaraguan peasant named
Oscar, who would have more in common with the biblical laborers

than most Americans, said, "I don't think the boss was unfair, because he didn't care about the work, or the profits it would bring. What he wanted was for everybody to be working."[7]

So, here we have this landowner going back to the market. Maybe he was just running some errands. (I am retrojecting my twenty-first-century sense of purpose for being back in the market.) But it does not matter why the landowner went. This character is the owner of the vineyard, and this character gets to decide whether or not to hire more workers. The landowner saw people idle, and the reason did not matter. The landowner thought, "I can put them to work," and hired them.

Another Nicaraguan who also has more in common with the laborers than we do, Felipe, said, "Here [Jesus is] saying that the kingdom of heaven is like a great farm, but a farm on which everybody earns the same so nobody will feel he's more than anybody else; people aren't separated by wages."[8]

Felipe might not articulate it this way, but his interpretation invokes *imago Dei*, this notion that humanity is made in God's image, best summarized by Genesis 1:26-27, "Let us make humanity in our image...So God created humanity in [God's] image." Or, this could bring to mind the equity in Paul's theology, evidenced in Galatians 3:28: "There is no longer Jew or Greek...slave or free...male and female; for all of you are one in Christ Jesus."

In the evening, the landowner called the laborers together and paid each of them a denarius. It was not a radical act of generosity. It is a subsistence wage. So, the people could afford to eat that day. The landowner paid each person a *dik'-ah-yos*, which means an equitable or righteous amount. The ones who worked all day thought the pay scale was unfair.

Have you ever had someone do something against you? You want justice. That is natural. You want what is fair. I understand. The problem is God plays by a different set of rules. We want restitution, but God offers transformation. We want what we are due, but God reminds us that we are all sinners in need of grace and are due nothing. The landowner reminds those who worked all day, "Friend, I am doing you no wrong; did you not agree with me for the usual daily wage? Am I not allowed to do what I choose with what belongs to me? Or are you envious because I am generous?"

Oscar pointed out that the farmer might have wanted everyone to be working. Felipe said that the story reminds us that we do not need to be separated by how much we earn. Maybe this passage is about God's free gift of salvation. Maybe it is about God's acceptance of people regardless of when in their lives they turn to God. Or, maybe this story is about God's desire for humanity to work together. John Hart suggests that this story describes the responsibility rich people have toward poor people.[9]

Regardless of the way we read the story, God's love is radical, and God does not play by human rules. If a young man wearing a swastika turns away from his white-supremacist friends and wishes to surrender his hatred, should the people around him hold him accountable for the hardness in his heart? If he committed crimes, then he will be held accountable for those. But this question is about hatred in the human heart. Taking off his swastika and turning away from a path that could only lead to violence and destruction must be an option if reconciliation can ever occur.

My friend Jerry mirrors God's attitude by forgiving a trouble-prone employee who struggles with addiction. The stories of this young man's escapades are many and sordid. Some of the other employees wonder why Jerry puts up with it. They wonder why he does not fire the young man after each misadventure. Yet, Jerry does no wrong to the other employees. He upholds his agreement with them. Is he not allowed to do what he chooses with what belongs to him?

We should not be jealous when other people experience God's love, even if we have known God longer and feel like we deserve a bigger portion. We do not. God's love is bigger than we are. God's love is transcendent. No matter where we are on our journey, we can cry out to God and know that our Lord hears our cry.

Notes

1. 2. Paul Tillich, *Systematic Theology, Volume 1: Reason and Revelation, Being and God* (Chicago: The University of Chicago Press, 1951), 213.

2. Friedrich Schleiermacher, *On Religion: Speeches to Its Cultured Despisers,* ed. Karl Ameriks and Desmond M. Clarke, trans. Richard Crouter, Cambridge Texts in the History of Philosophy (Cambridge: Cambridge University Press, 2003), 35.

3. Jon Sobrino, *Jesus the Liberator: A Historical-Theological Reading of Jesus of Nazareth*, trans. Paul Burns and Francis McDonagh (Maryknoll, NY: Orbis, 1993), 68–69.

4. Sobrino, *Jesus the Liberator*, 69, italics original.

5. Walter Brueggemann, "Exodus," in *New Interpreter's Bible*, ed. Leander Keck (Nashville: Abingdon, 1994), 793.

6. Sun Tzu, *The Art of War*, trans. Thomas Cleary (Boston: Shambhala, 1988).

7. Ernesto Cardenal, *The Gospel in Solentiname, Volume 3*, trans. Donald D. Walsh (Maryknoll, NY: Orbis, 1979), 180.

8. Cardenal, *Solentiname, Vol. 3*, 181.

9. John Hart, *Sacramental Commons: Christian Ecological Ethics* (New York: Rowman & Littlefield, 2006), 165.

PART THREE

Life

What is racism? What does a racist look like? When someone appears to be racist, how can those actively trying not to be racist engage or overcome differences? Is racism like hypocrisy? Everyone is a hypocrite; it is simply a matter of to what degree (see page 108). Is everyone a racist to varying degrees?

I have a Facebook "friend" who shares posts comparing black public figures with animals and was an active supporter of the birther conspiracy. Birthers doubt the legitimacy of Barack Obama's presidency because of a conspiracy theory that President Obama is not a natural-born U.S. citizen. Years ago, the friend and I were scuba diving buddies. We went diving often. We talked about life, where we had been, where we hoped to go, and had a lot of fun together. I am sure we talked about politics, but the conversations were not memorable.

After diving together for several years, my wife and I moved away. My dive buddy and I fell out of touch. Then, along came Facebook. A few years ago, we reconnected. Mostly, our reconnection was a walk down memory lane. Then, the racist-sounding posts started to appear. I was a bit confused until he shared a birther story. I approached his apparent racism by ignoring it.

Is it right to ignore racism? Ephesians 4:15 says, "Speaking the truth in love, we may grow up in all things into him who is the head, Christ." Sometimes it is easier to avoid a confrontation than to live the life of God's calling. Going around "speaking the truth in love" can be lonely business. To those who are at a different point in their faith journey, this truth-speaking can sound quite similar to condescension or holier-than-thou-speech. In this case, my former dive buddy does not profess a faith to guide his life.

A week after the events in Charlottesville, my former dive buddy sent me a Facebook message. He asked me to forward a political message on behalf of President Trump to all of my friends. In hindsight, I should have ignored it. As I have learned firsthand, email, text, Facebook messenger, and other text-only applications are not good places to engage in fruitful discourse. If I really wanted a relationship with my former dive buddy, I could have traveled to where he lived and sat down with him and had a conversation.

Trying to extend grace to myself, I could say that the timing of his message irked me. Charlottesville, the city where I had just moved with my family, was still roiling from the violence. For some, the turbidity created that weekend would last a long, long time. I could blame the move or the white supremacists' riot for my response. Or, I could own what I said. I choose the latter.

My response is my fault entirely. Instead of catching a plane to see my old friend or picking up the phone to call him, I responded with three hundred words on theodicy. I pushed back, not in love or empathy. I pushed back with a somewhat detached response ending with the question, "Do we stand against hatred, bigotry, and racism? Or, do we ignore it?"

The conversation went downhill from there. I tried and tried to bait him into denouncing racism. He refused. Only a few days before our conversation, the president defended the white nationalists who protested in Charlottesville, saying they included "some very fine people."[1] This was the moment, after a major event, when one response made sense: If you are standing in a line and realize that some of the other people in line are white supremacists, then it is up to you to get out of line. You cannot remain there and expect people to recognize that not everyone who stands with the white supremacists is a racist. Denouncing racism sounds easy. Just say it. Just acknowledge that it is wrong.

In response to my question about condemning the white supremacists, my friend brought up the 2016 U.S. presidential election, the Congressional Black Caucus, reparations, and the basis of wisdom. To me, it seems so simple. When one side has white supremacists, neo-Nazis, and KKK, we can say that side is wrong. This judgment does not exonerate counter protesters who behaved illegally. One can be against racism *and* liberal elitism, socialism, egalitarianism, and other left-leaning ideas.

Saying the white supremacists are wrong does not make a person a Democrat or Republican. It does not mean one is liberal or conservative, Christian or non-Christian, or anything else. White supremacy is an ideology that is antithetical to the Christian faith. Saying white supremacists are wrong does not mean Antifa or violent counter protesters are right. To me, the idea of condemning racism seems as though it should be so easy.

Yet, my friend could not say the white supremacists were wrong. It was as if we were playing some game and I did not know the rules. If he admitted they were wrong, it seemed as though he thought he would lose. Lose what? I do not know.

I became frustrated. How can we not condemn the white supremacists? Why is it so difficult to condemn white supremacy? Fear, frustration, dissatisfaction with life…Myriad factors form our worldview. Each experience contributes to who we are. My upbringing, social experience, work life, and education all contribute to my condemnation of white supremacy. What about my friend? I know some parts of his story, yet I still have trouble understanding how he cannot say white supremacy is wrong. Is he a racist? No one knows the human heart but God (1 Corinthians 2:11).

Dialogue is difficult work, but it is part of life and worth the effort. In the following two chapters, we will explore what brings people together and what pushes them apart. Both explorations are part of life and seek to overcome our differences.

Note
1. Rosie Gray, "Trump Defends White-Nationalist Protesters: 'Some Very Fine People on Both Sides,'" *The Atlantic*, August 15, 2017.

8
Bringing People Together

"Anticipate charity by preventing poverty, namely, to assist the reduced brother, either by a considerable gift or loan of money, or by teaching him a trade, or by putting him in the way of business, so that he may earn an honest livelihood and not be forced to the dreadful alternative of holding up his hand for charity..."[1]

—Moses Maimonides

"So then you are no longer strangers and foreigners, but you are fellow citizens with the saints and of the household of God."

—Ephesians 2:19

"Give a person a fish and you feed that one for a day. Teach a person to fish and you feed that one for a lifetime."

—Chinese proverb

What brings people together? Need? Once a man asked if he could wash his laundry in the church. I said, "No."

Then, I struggled with it for weeks. He approached a number of people as they were leaving the church on a Sunday morning. He is homeless and has asked for help on previous occasions. One time, we made an exception to help him. We allowed him to take a shower in the church under the condition that he would not ask us again but use the other ministries around the city that are set up to help with that particular need. So, when he asked if he could do his laundry, I said, "No."

Was I just being callous? Maybe. On that particular day, I was the last person in the church building. Another church has a laundry ministry. He could have washed his clothes at the other church. In Charlottesville, there is also a homeless ministry called The Haven, and it offers free use of washers and dryers. Still, I struggled because the Bible tells us to help those who need help.

In the Gospel of Matthew, Jesus talks about when we care for others, we care for him (25:31-46). In Hebrews, we find the instruction, "Do not neglect to show hospitality to strangers, for by doing that some have entertained angels without knowing it" (13:2). In Exodus, we read: "You shall not oppress a sojourner; you know the heart of a sojourner, for you were sojourners in the land of Egypt" (23:9). These three examples are only a few instances of biblical support for helping people.

Accepting Being Accepted

How can people overcome political difference when we ignore basic human needs? The man who asked for help needed help. I could judge the decisions he has made in his life to get to this point, but my judgment will not change his need for help. In this case, we divided ourselves based on haves and have-nots. I belong to the former category, and the man who asked for my help belongs to the latter. Yet, if I simply help him, then I am doing nothing for him in the longer term. Following the proverb, I would be giving him a fish instead of teaching him to fish.

In the faith journey, there is a sense of joining together with outsiders. Becoming one in Christ means reconciling with those with whom I disagree. It means overcoming differences and finding common ground in Christ. With a biblical edict to help people, how could I turn someone away? Were my afternoon plans so important that I could not wait with this man who is struggling in life while we washed his laundry? What about my old dive buddy? Are my goals and sense of what is right so much more important than his? I wanted him to admit that white supremacists were racist. For whatever reason, he feared making this admission. I could accept him where he was even if he could not accept me.

Jesus provides the ultimate example of accepting people where they are. His work on the cross and through his resurrection provide the basis for our relationship with God. We must accept that God accepts us. God has the freedom to make this choice and chooses to release us from the bondage of our sin. Why? Simply because God decides to do so. We accept being accepted.[2] If I accept being accepted, could I not accept my friend and let him decide whether or not to accept being accepted? Likewise, could I not accept the homeless man's frustration with my "no" and accept him as a brother in Christ, even if he is not at a place to accept being accepted? The shift between "we" and "you" language in Ephesians 2 points toward accepting being accepted. "You" is a directive. "We" joins the one who is directed with the speaker.

In Ephesians 2:10, it says, "We are what [God] has made us." In 2:11, the tone moves to "you," as in, "You were Gentiles by birth." The passage juxtaposes "once you were far" with the notion that "now you are near." The theology bears down on the mess we have created. The church is a worldwide institution with factions that are too numerous to name. We find unity in Christ and then find that we separate from others in Christ.

Separated from God or United with God

Ephesians 2:1-10 is about the shift from separation from God to union with God. How can this be a model for bringing people together in a politicized world? Verses 11-13 look at how we were once without Christ, but are now in Christ. This applies to Christ followers, but it might be harder to apply with people who do not profess faith in Christ.

Even in the original context of Ephesians 2, there were divisions. Jewish people and the larger Greco-Roman society experienced tension generally, and among those who shifted to follow Christ, there was religious tension. The original recipients of Ephesians would have found it shocking to bring up this tension. It is almost as if they were just getting over their differences, and then Ephesians 2 brings it up again. The weight of hearing the distinction without returning to the separation would rest on the phrase "you once were" (Ephesians 2:2).[3]

The church is the result of God's reconciling work in Jesus Christ. For both Christians and non-Christians, the call to unity can be a model for overcoming differences. In Ephesians, we read that the perfect church, the church of God, is not full of fractious divisions. This church, the unified church of God, is aspirational. It probably never existed. But twenty-first-century people can work at being more unified. Just as we can never be perfect in our faith journeys, we can continue to grow throughout our lives.

All of the faith divisions of the twenty-first century lead us to build walls that separate us from one another. Christians isolate themselves from other Christians. However, Christ breaks down these walls between us.[4] Sometimes the divisions between us seem to represent something good. Houses have walls and a roof to protect furniture and occupants from the rain. It also helps keep honest people honest when we lock our possessions away from our neighbors. This seems necessary, because some people are not honest. Without locked doors, some people might take what we have or try to hurt us. These divisions and barriers between us can seem good.

Other divisions are not good. Our perception of divisions that separate us from the other are bad. The divisions that make us see people differently are bad. When someone lives in a particular neighborhood, when someone has a bumper sticker reflecting a particular ideology or political belief and denigrates others, when someone has a different socioeconomic status from our own—we can see these divisions as bad.

Some churches seem as if they should be separate from us. They say things that misrepresent God, the Bible, and the Christian tradition. Having the freedom to disassociate from those kinds of churches and the division between us gives us space to grow. Before getting too far into critiquing the theology of other churches, each one of us must be careful about the planks in our own eyes. Our beliefs can have some huge shortcomings. What bothers me might not bother you. What forms a barrier between me and God might not be an issue for you. The message of Jesus Christ is in the middle of these divisions.

When we experience God's transformation, we can set aside differences and come together in Christ. No longer do we hang our hats on this specific belief or that belief. Instead, we find wholeness in Christ. Joining other Christians in doing God's work, we find unity. We do ministry with our neighbors instead of ministry to the other.

What about the man who asked if he could wash his laundry in the church? The man was looking for someone who carried out a ministry-to. A ministry-to is one where some people serve and other people receive their service. Instead of building a relationship and ministering with one another, a ministry-to moves in one direction. It moves from giver to receiver. It moves from helper to person-being-helped. The man seemed to want to remain the other. But God calls us to do ministry with. I could learn from him. I could grow with him. The basis of his salvation is the same as mine. God has already chosen to accept him in his sinfulness, just as God accepts me. Does he accept his acceptedness? That is between him and God.

Joining together in ministry carries risk. We have to open ourselves. To enter the faith journey as partners means, for me (and
us), recognizing that we do not know everything. Unity in Christ
means putting God first, ahead of the self. The challenge is to resit-
uate the conversation. He asks, "Can I do my laundry in your
church?" At that moment, my calling is to love him where he is,
just as God loves me where I am. Society's challenge is to love him
where he is, show him hospitality, and invite him to participate.

He might ask, "Does this mean I can always do my laundry in your
church?" Maybe. Maybe not. The answer is more dependent on the
leading of the Holy Spirit than a deterministic theology of do this or
do that. When we engage, we can "put on Christ Jesus" (Romans
13:14). We are on a journey. We have the opportunity to grow each
day. We each have the choice of whether or not to take advantage of
the opportunity and "put on Christ Jesus," put God first ahead of the
self, and to seek unity in Christ. We are all one in Christ.

Bridging Gaps: The Life of Kevin Ly

What kind of person brings people together? Up to this point, most
of the examples have been about people being driven apart. At the
Unite the Right rally in Charlottesville, division played the central
part. Each side was opposed to the other. My former dive buddy
pushed a conspiratorial worldview that fosters divisiveness. The
homeless man stretched notions of ministry and forced the question to either help him or not help him. These encounters carried
some level of confrontation. For inspiration, I look to people who
bring others together, people like Kevin Ly.

I met Kevin Ly in the 1990s. He was a teenager in my brother's
first pastorate. Though his outward life bore the marks of a committed Christian, he would not make a public profession of his
faith. He would not seek baptism. He would not confine his faith
to a label or a ritual. The reason was simple. He respected his par-

ents. He lived the commandment to "honor your father and mother" (Exodus 20:12). His parents were refugees from China and would not permit their son to convert to this American religion. So, he did not.

As a young man, leading the youth of Greenbrier Baptist Church, he had wisdom beyond his years. He had patience for unruly teens when I had none. He knew Bible stories. He understood a faith that was both simple and all-encompassing. And, even though he had not been baptized, he helped lead some of the young people to make a profession of faith and seek baptism.

After we moved away, Kevin continued his faith journey. Eventually, his parents allowed (blessed?) his conversion to Christianity. He made a profession of faith and experienced the baptismal waters. Then, he attended seminary. A church ordained him into ministry.

Kevin and I crossed paths again in 2015. Seeing him rejuvenated my soul. We ran into each other at a denominational event. Those events can be draining. Sometimes they represent the worst of the institutional church. I walked out of a session wondering why God trusted the church to humanity. Then I saw Kevin. His smile brightened my day, and he reminded me of why we do what we do. He wanted to serve God. I felt God blessing me with the opportunity to encourage him. I do not know if it meant anything to him, but seeing him meant much to me.

When Main Street Baptist Church in Luray, Virginia, called Kevin to be their pastor in 2016, I went to his installation service. He asked me to participate, and I felt honored to do so. When I started my new ministry at University Baptist Church in Charlottesville, Virginia, in 2017, I asked Kevin to participate in my installation service. He said, "Hey Matt, but of course! Am honored to be invited."

That response was Kevin. He had enthusiasm for life and it was contagious. His zest for ministry, food, friends, games, and fun was impossible to avoid. His response made me more excited about my own installation service. His infectious smile caused everyone around him to smile.

Kevin had a unique gift of drawing people together. His unassuming approach did not mask weakness. He was an athlete and loved to play games. His sense of humor disarmed pretense and his wit lightened almost any mood. In October 2017, Kevin and his fiancée, Jackie, celebrated their marriage. Six weeks later, Kevin had a massive heart attack and died instantly. He was thirty-eight years old.

To think I will not see Kevin anymore in this life is difficult. Why would someone so vivacious be snatched away? Why would God allow this to happen? Why would someone who can bridge the gaps in twenty-first-century U.S. society drop dead while others who foster division keep pushing people apart?

The truth is, we do not know the answer. Did God take Kevin? No. Did God cause this to happen? No.

In 1 Corinthians 2:9, Paul quotes Isaiah 64:4: "What no eye has seen, nor ear heard, nor the human heart conceived, what God has prepared for those who love him." Then Paul adds his commentary in verses 10-11, "These things God has revealed to us through the Spirit; for the Spirit searches everything, even the depths of God. For what human being knows what is truly human except the human spirit that is within? So also no one comprehends what is truly God's except the Spirit of God."

Life is a mystery. Every person will die. When someone dies young, we mourn in a special way. We mourn the loss of a life not yet fulfilled. We mourn the potential, or what could have been. In Kevin's death, we can ask why someone so good would go so soon? We can ask, but we may never know. What caused his body to fail him? Doctors might figure out what happened, but it will not bring him back.

People like Kevin can inspire us to live better. If Matthew 25:31-26 points to the nature of God, then how we interact with other people has eternal significance. This does not mean salvation is based on what we do. Instead, a person's relationship with God is reflected in how that person lives (James 2:17). People like Kevin can bring others together.

Turn and Live

Fighting for control seems to be the nature of humanity. In a micro-cosm of world affairs, we do it in our lives. We all want control. On a global stage, nothing seems to change. Current headlines are full of nations struggling for control over who gets to decide what about another country's national interest. Looking backward, this theme is remarkably consistent. Throughout history, major world powers have fought for control.

Twenty-six to twenty-nine hundred years ago, Assyria, Babylonia, and Egypt maneuvered to control the region. In 605 BCE, Babylonia gained the upper hand. Focusing on Judah and the Jewish people, when Babylonia beat Egypt in a decisive victory, Judah's fall was inevitable. For the Jews, this was inconceivable. How would God let them fall? But, when Judah fell, King Nebuchadnezzar II took Judah's best and brightest back to Babylonia. Ezekiel was part of this brain drain, and the prophecy under his name in the Bible sheds light on the nature of God in rela-tion to bringing people together.

Before looking at how Ezekiel brings people together, we must explore control. How does control work? When we have it, we like it and want to keep it. When we lose it, we want it back. Ezekiel refocused this thirst for control. Instead of control, he focused on God. His visions contrasted with the status quo of his day. He spoke against Judah and against Jerusalem, not because he had a problem with the physical location of Judaism, but because the place had taken precedence over God.

Speaking against Jerusalem would have been an incredibly unpopular prophecy. The captive people wanted to go home. They wanted their army to rise up and defeat the Babylonians, but God wanted their commitment. God wanted them to surrender control to God. Instead, they wanted control for themselves.

Many people struggle with surrendering to God today. In some ways, the reason Kevin could bring people together was his strength in his weakness. They grow up in a prominent Christian home, have parents who are leaders in the church, and find powerful mentors. Kevin had none of those advantages. He did not have influential parents within his Baptist tradition or leading mentors. He had his faith, and his faith made him strong.

One of the biggest obstacles my former dive buddy encountered was himself. He would have to surrender his will to God's will to become a follower of Christ. The definition is in the name: Christ-follower. Yet he wanted to do things his way.

The homeless man who asked for help did not want to hear about transformation. He wanted a transaction. In other words, he viewed our encounter as an exchange. He seemed to see me and the church as a group of do-gooders. Thus, we need someone to whom we can do good. He was fulfilling his role as the recipient of good deeds. All would be well. Except, transformation interrupts transactional faith.

Taking Responsibility

In Ezekiel 18, the people did not want to take responsibility for their own actions. They did not want to admit they wrestled control of their lives from God. Instead, they wanted to blame someone else. They blamed their parents or their parents' parents. To put it in ancient terms, they said, "Who of our ancestors sinned that we now suffer?" (18:25).

Today, people say, "It's not my fault! It's society's fault! It's my upbringing. It's someone else's fault."

The problem with this idea of transgenerational punishment is people's reluctance to accept responsibility for their actions. Ezekiel makes it clear that life's unfairness is not related to God. We cannot blame God when things go wrong. We cannot blame our parents, nor can we blame anyone else when we make bad choices. Each one of us makes choices. Some are good. Some are bad. When we make a choice, we have to live with the outcome. Ezekiel pushes against this kind of blame game.

For Ezekiel's audience, the ones who suffered under King Nebuchadnezzar II, the geopolitical landscape might not have included a viable option for them to avoid foreign occupation. They truly suffered. Ezekiel provides a commentary on reality and a correction to the fallacious belief that God caused someone's suffering. It was not God's fault. King Nebuchadnezzar II and the Babylonians attacked and overtook them, not God. Judah was not a major power. If it had not been Babylonia that conquered them, it might have been Egypt.

For Ezekiel, the answer to this question about occupiers in their land and some people being exiled is God. The answer is not returning to Jerusalem or the religious status quo. For us, too, the answer is God, not finding someone to blame. God does not call Christians to go back to the way things used to be. It seems like human nature to look back at some moment in a glorified past or an imaginary, wonderful past. It would be alluring to long for that time. Yet, God is at work in the present. God will continue to be at work in the future, and God wants us to be involved in that work and transformed.

Ezekiel disconnects the suffering that surrounds him from previous generations. For us, the message is what we do matters. If we blame others for everything that goes wrong, we miss opportunities to grow. We can connect this message with New Testament readings, like Hebrews 13:16: "Do not neglect to do good and to share what you have, for such sacrifices are pleasing to God." Or, we see in Jesus' words in Matthew 21:32: "John came to you in the way of righteousness and you did not believe him, but the tax collectors and prostitutes believed him." We are free to accept or reject God. God's way is fair. God is not punishing us.

Regarding Ezekiel 18, Ellen F. Davis writes: "God's justice is connected with the radical affirmation of the human freedom to change (repent) and also with divine pathos, God's desire for Israel's repentance."[5] God has compassion for humanity. In this case, God's prophet Ezekiel is delivering the difficult message of repentance from our own attempt to connect God and fatalism. In other words, for Ezekiel's audience, they might have said, "Oh! I cannot do anything because God wants me to suffer." However, God does not want us to suffer, but to be free.

For his listeners, the saying about "parents eating sour grapes" (Ezekiel 18:2) let the people off the hook. The full verse is: "What do you mean, that you use this proverb concerning the land of Israel, 'The parents have eaten sour grapes, and the children's teeth are set on edge'?" Instead of being transformed, they could blame God for their problems. They could say, "Oh no! The parents have eaten sour grapes!" God offers all of us the possibility of life. This is a possibility that is available despite suffering throughout history and human sinfulness.[6] We have a choice, but we must make our choice. God cares about our choices.

The choices we make can either bring people together or push them apart. We can enter into dialogue, listen to the other, and try to empathize with the other's perspective. Or, the divisions pervading the U.S. can continue. Either/or language might seem stark. Søren Kierkegaard writes: "Yes, now I see it all perfectly; there are two possible situations—one can do either this or that. My honest opinion and my friendly advice is this: Do it or do not do it—you will regret both."[7]

Regardless of what we do, some situations feel as if they have no positive outcome. Engaging in dialogue with a person who holds oppositional views takes work. It can be tiring, and these discussions can feel pointless. Not only might someone like my old dive buddy not change their perspective, but he could confront me with truths about myself that I am not prepared to address. I would regret the dialogue and have overcome no divisions.

Likewise, doing nothing allows divisions to continue. Watching the polarized world continue to go by is regrettable. Kierkegaard is right about potential regret in both situations. However, regret is not quite so deterministic as it might sound. If it were, the logical response would be to take the choice that carries the least cost. Obviously, doing nothing usually has little risk. Yet, we do not know that both choices are regrettable. Therefore, the solution is to do something and try to bring people together.

Notes

1. Moses Maimonides, *The Wisdom of Moses Maimonides* (White Plains, NY: Peter Pauper Press, 1963), 10.

2. Paul Tillich, *Systematic Theology, Volume 2: Existence and the Christ* (Chicago: The University of Chicago Press, 1957), 179.

3. Pheme Perkins, "Ephesians," in *New Interpreter's Bible*, ed. Leander Keck (Nashville: Abingdon, 2000), 404.

4. Kevin Baker, "Wrecking Crew," *The Christian Century* 123, no. 14 (2006): 21.

5. Ellen F. Davis, "Ezekiel 18 and the Rhetoric of Moral Discourse," Gordon H. Matties, *The Jewish Quarterly Review* 84, no. 4 (1994): 502.

6. Joseph Blenkinsopp, *Ezekiel*, ed. James L. Mays, Patrick D. Miller, Jr., and Paul J. Achtemeier, Interpretation (Louisville: John Knox Press, 1990), 84.

7. Søren Kierkegaard, *Either/Or Part II*, ed. Edna H. Hong and Howard V. Hong, trans. Edna H. Hong and Howard V. Hong, vol. IV, Kierkegaard's Writings (Princeton: Princeton University Press, 2013), 159.

9

The Lines We Draw

"We weaken our greatness when we confuse our patriotism with tribal rivalries that have sown resentment and hatred and violence in all the corners of the globe."[1]

—John McCain

And now I will tell you
what I will do to my vineyard.
I will remove its hedge,
and it shall be devoured;
I will break down its wall,
and it shall be trampled down.
I will make it a waste;
it shall not be pruned or hoed,
and it shall be overgrown with briers and thorns;
I will also command the clouds
that they rain no rain upon it.

—Isaiah 5:5-6

The Red Queen said to Alice, "Do you know Languages? What's the French for fiddle-de-dee?" "Fiddle-de-dee's not English," Alice replied gravely. "Who ever said it was?" said the Red Queen. Alice thought she saw a way out of the difficulty this time. "If you'll tell me what language 'fiddle-de-dee' is, I'll tell you the French for it!"[2]

—Lewis Carroll

People never cease to separate themselves from one another. Even in close relationships, there are limits to how much people share. These limits form boundaries. Moving beyond these boundaries, there can be a sense of violation. To better understand the lines people draw between themselves, we can look in the Bible and see the way people both separate themselves from God and move beyond the boundaries to reunite with God. The way people relate to God can be helpful in exploring the way we relate to one another.

How often does life hang by a tenuous thread? How many friendships have become fractured by something that seems small? A church votes to become inclusive and welcome all people regardless of their sexual orientation, and some people leave the church because of the vote. All of the other things that made the church a spiritual home for so many years still remain. In this example, the church has made one significant change. But the attributes that define the church still remain.

In previous generations, churches voted to welcome women into leadership. When churches made that move, some people left. Going back further, when white congregations voted to welcome African Americans, some people left. Almost every theological shift involves wrestling with people who feel differently than those who decide to make the shift. Changes inevitably leave some people out.

When someone holds to a belief about God, that belief can provide the ground of being. For every Christian, what we believe about God undergirds the experiences in life. Every Christian, to some extent, draws lines based on those beliefs. If, for example, belief in God includes belief that God made only male and female to be in a marital relationship, then belief compels that person to draw lines excluding same-sex couples and, possibly, those who accept same-sex couples. For people with that theology, God requires no less than their dogmatic practice.

A counterpoint to this dogmatic adherence to boundary keeping would be Jesus' various critical responses to legalism. According to Mark, for example, when a scribe asks Jesus which commandment was greatest, he answers: "Love the Lord your God with all your heart, soul, mind, and strength. The second is this: Love your neighbor as yourself" (Mark 12:28-31). People who try to follow God can view Scripture and contemporary issues through the lens of Jesus Christ.

Isaiah's Inclusive Poetry

When we look at the Bible, we have various ways of trying to understand and apply it to our lives. Some passages are trickier than others. Unfortunately, some of the ones that appear to be straightforward are sometimes saying something different than we think. For example, consider the Book of Isaiah. It was composed over a long period. The prophecy covers the reigns of the four kings listed in the opening verse. Then, chapters 40–55 encompass the Babylonian exile, and the final eleven chapters focus on temple matters.[3] In each, the book speaks to a current situation in the lives of God's people.

We can understand the prophetic tradition of Isaiah speaking to the current lives of the ancient Israelites, but Isaiah can also speak to a hurting and struggling world. Still, applying an ancient text to today's problems can be difficult. Deciphering the meaning for today can be a bit like Alice's conversation with the Red Queen. We do not know exactly what the text means.

One method of unlocking a text's meaning is structural criticism. The philosophical anthropologist Claude Lévi-Strauss first applied this literary technique to reading the Bible.[4] This approach suggests that we focus on the text as it stands, not on authors or editors or how Isaiah came to be, over how long, and who really wrote it. The meaning is what it conveys to the readers. Contemporary readers can ask about the meaning for the twenty-first-century U.S. So, we read a passage, like Isaiah 5 and apply it to our lives.

Isaiah 5:7 says: "For the vineyard of the LORD of armies is the house of Israel, and the people of Judah are his pleasant plant: and the LORD looked for justice, but, behold, oppression; for righteousness, but, behold, a cry of distress." This verse summarizes the meaning of the section: God wants justice justice (*mish-pawt'*), but we have bloodshed (*mis-pawkh'*). God wants righteousness (*tsed-aw-kaw'*) but we hear cries (*tsah-ak-aw'*). The Hebrew text shows the similarities between words for "justice" and "bloodshed" and "righteousness" and "cries." In each case, the difference is one letter.

Isaiah 5 is poetry. "Let me sing for my beloved" (5:1). Like other poetry, there are spaces between the lines that the listener must fill. As is common in Hebrew poetry, the vineyard represents the beloved. The owner of the vineyard did everything he could to promote growth. He dug the vineyard, set it up, and prepared it. Anyone who gardens knows that a lot of work precedes planting the first seed. When the harvest came, instead of producing grapes for wine, the vineyard has "wild grapes," (*be-oo-sheem'*), which could also be translated as "poison berries."

The poem shifts. The prophet turns to the audience and asks them to "judge between me and my vineyard" (Isaiah 5:3). Who is guilty? Who failed? The owner assumes that the verdict is against the vineyard and moves to judgment. He will return the land—this land that he carefully cultivated—to waste. What started as a love song has now become a trial. Isaiah speaks for the vineyard owner and argues a case before the Israelite audience.

What does Isaiah say about justice and righteousness? Political divisions are not limited to same-sex marriage and racism. As a nation, we refuse to engage in rational conversations about responses to mass violence. The knee-jerk reactions are (a) reduce the number of guns or (b) have more people with more guns. I confess to being in the former category. I cannot understand how the presence of more guns could help solve the problem of mass violence. After each mass shooting, I struggle with what to say. Churches can organize prayer services, although I cannot even remember my first one. After a mass shooting, we can certainly pray, but God expects justice (*mish-pawt'*). Instead, we have bloodshed (*mis-pawkh'*). God expects righteousness (*tsed-aw-kaw'*), but we hear cries (*tsah-ak-aw'*). In Isaiah 5, the song of the unfruitful vineyard shows what God wants.

A parallel to Isaiah's speech on behalf of the vineyard owner is 2 Samuel 12. Nathan the prophet goes to David the king and tells him a story. The story is about a rich man who stole a poor man's little lamb. David is outraged against this criminal and pronounces judgment. Nathan says, "You are that man!" David took Bathsheba and had her husband Uriah killed. In Isaiah 5, we see the same indictment against Isaiah's audience. *You are the guilty ones!* Thank goodness Isaiah is not saying that to us. We can use Lévi-Strauss's structural criticism, describe this passage as poetry, and keep it at an arm's length.

According to Gene Tucker, "The first rule of biblical interpretation should be this: Do not reverse the miracle at Cana"[5] (John 2). In other words, do not turn the wine into water. When interpreting Isaiah, we cannot replace the words themselves. Isaiah speaks to a people who have displaced the love for God and one another. God is not the center of their lives, so the vineyard faces judgment. This failure of justice and righteousness is a frequent theme in prophetic writing. Often, when justice fails, the powerful have taken advantage of the weak. Instead of keeping Isaiah at arm's length and away from the politicization of twenty-first-century United States, we can let this passage speak to us. We can ask ourselves if we have any role in the failure of justice in our world.

Have we heard cries when we should have been righteous? Do we ignore the stories of women who have been victims of sexual violence? Do we grow quiet after a mass shooting because we would rather not confront gun rights activists? Do we look past racial and sexual micro-aggressions and hope things will get better?

Putting together prayer services is easy, but how can we call for prayer services when we just keep killing each other? How can we hear God and live out the kind of justice God wants for our world? Gun violence, racism, and sexual assault are not political issues— the Christian life is about God and addressing the issues of life through the lens of Jesus Christ. What about the vineyard of our lives? Are we producing grapes for the wine press? Or are we producing poison berries?

Perhaps we cannot change the world, but we can change ourselves. When Alice thought of a clever response to the Red Queen in Lewis Carroll's story, the Queen "drew herself up rather stiffly, and said, 'Queens never make bargains.'"[6] Do we find our place in God's story? We can hear Isaiah and say that we are not trying to make a bargain with God. We want to make a change in ourselves. We want to produce God's justice and righteousness. We can share God's love and mercy and erase the lines we draw.

God's Possessions Break Barriers

In the Gospel of Matthew, some Pharisees plotted to entrap Jesus. They tried to set him up with false praise: "Teacher, we know that you are honest." After trying to build him up, they asked about paying taxes. Jesus could see through their trap and called out their deceitful motivations. Then, we read the famous biblical aphorism, "Give to Caesar what is Caesar's" (Matthew 22:15-22).

This passage falls after Jesus' triumphal entry into Jerusalem (Matthew 21). He cleanses the temple, curses a fig tree, and answers questions from the chief priests and elders. He tells the parable of the man with two sons. Then, he tells more parables—the one about the wicked people who rented a vineyard and a cryptic one about a wedding banquet. Jesus draws lines of true commitment to God and commitment to human creations.

In this passage, some Pharisees team up with Herodians. This is an unlikely duo. Pharisees, in principle, resisted paying this particular tax, whereas the Herodians supported the Roman regime and paying this tax. Yet, these two groups find a common enemy in Jesus.

For the Pharisees, Jesus represents something different from their view of institutional religion. They liked their power and avoided demonstrations because any such protest would give the Roman powers an excuse for savage retaliations. Pharisees had a sort of truce with the Roman oppressors. The Romans did not do anything too outrageous, like trying to put up a picture of the emperor in the temple, and the Pharisees made sure their institutional religion did not impede the Romans.[7] The Herodians were wealthy aristocrats who maintained friendship with Herod Antipas.[8]

The tax in this text is not abstract. It is a specific Roman tax on the subjugated people. We take this story and apply it generally, but it is specific. According to Eugene Boring, it is "not an instruction on how people who live in a complex world of competing loyalties may determine what belongs to Caesar and what belongs to God."[9] This tax had particular meaning for Pharisees, Herodians, Jesus, and anyone listening. It also had meaning for the original hearers of the gospel forty to fifty years later.

Although the lesson is not abstract, there is a distinction between what belongs to Caesar and what belongs to God. Certain lines are Caesar's and others belong to God. Focusing on the Greatest Commandment, God's lines include and draw people together. Leonardo Boff writes, "Authority is a mere function of service."[10] Caesar had a function in the ancient Greco-Roman world. Politicians have a function in today's world. Their authority extends as far as their service and role. Beyond that, their authority ends.

Jesus is free from preconceived ideas, like these human categories of Pharisee or Herodian. Instead of answering directly, he changes the question. He will not be trapped in the world of human knowledge and understanding. We frame issues in a linear fashion: liberal or conservative, this party or that party, pro or against this issue or that issue. Jesus reframes the paradigm. He focuses on humanity and God. To understand his response, we must better understand both the tax and what happened between this conversation and when the Gospel of Matthew was written.

The tax refers to the "census," (kane'-sos), a Roman head tax instituted in 6 CE. This was when Judea became a Roman province. By the time Jesus, the Herodians, and Pharisees had a conversation, the tension surrounding this tax was mounting. If Jesus agreed with the Herodians and supported the tax, his Jewish followers would lose faith in him. They would see him as complicit with the oppressors. If he agreed with the Pharisees and opposed the tax, the Jews would be pleased, but the Romans would have sufficient evidence to charge him with sedition. The religious leaders are trying to trap him. Their sarcastic words of praise do not fool him: "Teacher, we know that you are, *al-ay-thace'* [English: true or honest]." The NRSV says, "sincere."

This was a divisive issue. Public opinion varied and people held deep feelings. This *kane'-sos* tax triggered the nationalism that grew into the Zealot movement. Some people even argue that Jesus was part of the Zealot movement,[11] although texts like Matthew 22:15-22 undermine that position.[12]

Sometimes the Bible can seem distant, as though it describes a place far, far away and people we do not know. In our safety and security, many of us cannot imagine what it was like to have a foreign oppressor inscribing our lives with their tax. Yet, for the people in the ancient world and many who have heard these words over the centuries, faith is a matter of life or death. Part of the willingness to give to God what belongs to God goes with counting the cost. What does following Christ cost us? If the answer is nothing, then we are doing it wrong.

Jesus lived a sacrificial life and calls us to follow him. The Gospel of Matthew was probably written between 80–90 CE, which means it came after the First Jewish-Roman War. This means that Matthew's community not only knew about the crucifixion and resurrection, but they knew where this story was headed. The conflict over this *kane'-sos* tax was not going away. Jesus' answer mattered. Rereading the story in Matthew 22:15-22, we can sense his irritation when he sees through their pleasantries and calls them out as hypocrites.

Every one of us is a hypocrite—it is just a matter of degree. We split hairs with Jesus in our lives all the time. We essentially say, *I will give you this, but not that, God.* Or, we recognize God's inclusiveness, then draw boundaries for our faith. Why do we hold back from God? We hold back because we are afraid of what we might lose. Jesus does not follow the same playbook as his enemies. He follows God. According to Paul Fiddes, this gives him "outrageous freedom in matters of religious law, so he allowed nothing to qualify God's claim upon the whole of life."[13] We can give Caesar his due, yet there are limits. We can only "render unto Caesar" if it does not conflict with what God demands of us.

Jesus did not have the coin for the *kane'-sos* tax, or the Bible does not tell us that he did. He asks someone to show it to him. The Bible does not say that he touched it. They could only pay this tax with a Roman coin, and each coin had an image and inscription that many Jews considered blasphemous. It read: *Tiberius Caesar Divi Augusti Filius Augustus Pontifex Maximus* ("Tiberius Caesar, august son of the divine Augustus, high priest"). Even to hold the coin meant to participate in a system that blasphemed their faith. Jesus veered close to human-defined lines, but he did not cross them.

The Pharisees and Herodians talk about "paying" the *kane'-sos* tax. Jesus' reply uses a different Greek verb. He says, "Give to the emperor those things that are the emperor's." The crucial moment of this entire passage is the next phrase: "and to God the things that are God's." For us to understand God and our calling, Gustavo Gutiérrez writes that we must "eradicate all dependency on money"[14] and cultivate a sacrificial life. What is God calling us to give up? What does Jesus want us to return to Caesar? What boundaries does God want us to break down? What lines does God want us to erase?

Working for God

When I was a young, I visited my aunt and uncle who lived in Ewing, New Jersey, near Trenton. My uncle took me to the Quarry Swim Club in Hopewell. There was a high ledge where people could jump into the quarry. I was probably twelve at the time, and he allowed me to go off the high jump. It seemed like it was two hundred feet high, but it might have been twenty feet. I remember that feeling of putting my toes over the edge and thinking about turning back.

I trusted my uncle. He said it was safe. Yet, moving over the edge of that diving board seemed impossible. My adolescent fear created a line that did not need to exist. What was at stake? Truthfully, not much of eternal significance. However, the experience shows how artificial boundaries can prevent the abundant life Jesus speaks about in John 10:10. He said, "I came that they may have life, and they may have it abundantly." Jesus talks about the people who try to enter the sheepfold by climbing over the fence instead of coming through the gate. He says he is the gate and those who enter through him will find a safe pasture. This is the abundant life of John 10:10, and those who enter through him can experience it. Those who try to find an abundant life without Christ miss some of the richness of experiencing it with him.

Christians regularly stand with our toes over the edge, ready to jump into the future—knowing that we jump in faith. God is with us, leading and guiding. How can we be agents of reconciliation when we draw lines that prevent us from moving beyond our comfort zones?

Recall the experience in Charlottesville I shared earlier. If I had been unable to move between the neo-Nazi and counter protestor, I would not have seen their humanity. They would both remain nameless, faceless symbols, and not human beings created in the image of God. Their behavior is still abhorrent, and it is inconsistent with a biblical worldview, but, the two men are children of God.

Throughout the Bible, there are familiar stories. They explore moral behavior, Christian ethics, and justice, not once, but repeatedly. Christians gather each week and revisit topics, not because we are incapable of remembering or understanding them. We study the Bible and worship, because we need to hear the message again. We are not Pharisees who live by a list of rules, where we memorize that list and then check each action, each day against that list. We are transformed in Christ into new beings. Transformation is a constant journey. It never ends.

Transformation allows Christians to approach the boundaries that people create and react to each one. Life is not like a person following the Ikea directions, in which one simply needs to follow steps in order in order to achieve an expected outcome. Living is about transformation. The biblical stories, lessons, songs, and prayers are part of the journey. They are about Jesus Christ, not a random historical character named Jesus. As followers of Christ, we do not live in nostalgia, but celebrate Jesus Christ in the present.

Paul Tillich writes: "Without the manifestation of God in humanity, the question of God and faith in God are not possible. There is no faith without participation."[15] Our Christian lives involve action. None of us are along for the ride. We all have a part to play. In the faith journey, people are active participants. There are no bystanders. In 1 Thessalonians 2:9, the apostle Paul says, "We worked night and day...you are *mar'-too-rhes*." The Greek means witnesses; by analogy, it means a "martyr." Encountering Christ changes us. We participate, live out our faith, come to worship, experience transformation, and reflect Christ.

Life in faith is a radical journey of transformation. The earliest Gentile followers of Christ went from polytheism to monotheism, belief in one God. They did not just believe in any god, but they believed in the God of Israel who was also the creator of everything. They set aside everything they had believed and accepted Jesus Christ as the redeemer. With the Holy Spirit, they are a triune God—three in one and one in three.

For Paul, the gospel is not transactional. Instead of giving the good news in exchange for payment, Paul and his missionary partners shared God's grace for free. They worked as some sort of laborers to pay their way. Money did not cloud the message of the gospel. Liberated from the distraction of paying Paul's operational expenses, listeners could focus on hearing what he said. He spoke of God's grace and what the Christian life looks like. Unlike modern Christians, they did not have the joy of financially participating in God's work. That would come later as the fledgling church moved from the expectation that God would return very soon to an expectation that God would return someday. Until that day, Christ followers can live out their faith in the community called church.

God's church offers something special. The entire history of Christianity points to the church containing divine knowledge. Like the earliest Christ followers, twenty-first-century Christians participate in something special, something divine. The apostolic succession from Paul to the Gospels and the rest of the New Testament and to today is unbroken. There are certainly many strains in the Christian tradition. Some Christians hold fast to their dogmatic lines and do not speak to others. No one will see eye-to-eye with every sister or brother in Christ. Pope John Paul II writes,

"The knowledge which the church offers to [humanity] has its origin not in any speculation of her own word, however sublime, but in the word of God which she has received in faith."[16] What Christians share, sing about, and read is not our own, no matter how clever or wonderful it is. What we share, sing about, and read is divine. It is of God.

The good news that Paul preaches is nothing less than the revealing word of God working to bring about faith and keep it alive. The central theme of the apostolic preaching remains the message of the crucified Christ's resurrection. To twenty-first-century ears, gruesome talk of the crucifixion can confront contemporary sensibilities. It is part of the story. Gerald O'Collins describes it as "the already achieved climax of divine revelation to which believers look back and to which, in faith, they here and now continue to respond."[17]

We work for God when we respond, when we participate in the Christian life. But we have to be careful that our work does not replace God's work. Is our initiative part of the solution? Possibly. Start the conversation by recognizing that God is not confined by the Bible or our preconceived beliefs. When God resembles us, we have created God in our own image instead of being re-created in God's image.

Fyodor Dostoevsky's story of the Grand Inquisitor is about someone who let his work replace his calling in God. In the story, Jesus returns to Earth in bodily form and goes to Seville, Spain, in the sixteenth century, during the most terrible time of the Inquisition. Jesus walks among the people, and in his mercy, he performs miracles with a gentle smile and infinite compassion. He heals one person when they touch his garment, reminiscent of the woman in Mark 5. He gives sight to a blind man, like Bartimaeus in Mark 10. The people of Seville recognize Jesus, and they adore him.

The Grand Inquisitor has Jesus arrested and sentences him to be burned the following day. Then, the Grand Inquisitor visits Jesus in his cell. The Grand Inquisitpr is an old man, maybe 90 years old. He points his frail finger at Jesus and says, "Is it you? You? Don't answer; be silent. For what can you say? I know too well what you would say, and you have no right to add anything to what you said in the Bible. Why did you come to hinder us?"[18]

That is how the Grand Inquisitor sees Jesus—as a hindrance. Jesus gets in the way of the world the Grand Inquisitor has worked to create. Jesus gets in the way of social expectations. Jesus gets in the way of *our* work. The Grand Inquisitor would celebrate Christians coming together to worship in churches each week as long as Jesus Christ is not part of it.

There is something special in church, something divine. People's participation reflects their faith. They experience the transformation of Christ's presence in churches. Paul urges, encourages, and pleads with the recipients of his letters to live a life worthy of God. Today, we can read these earliest Christian words. We can make them our own. When we do, Paul speaks the same message to us.

Jumping Off the Ledge

John McCain served in Vietnam as a naval pilot. When the Vietnamese shot down his plane in 1967, they captured him and he remained a prisoner of war until 1973. He retired from the Navy in 1981 and entered politics. He served in the U.S. Senate from 1987 until his death in 2018. His political record is typical of many long-serving politicians. There are some high points and low points. In death, we should not elevate McCain's status or seek some sort of Protestant beatification.

McCain's final act was to write a letter to a nation that keeps dividing itself into more and more partisan groups. The scene he left in Washington is highly politicized. To that world, he wrote a letter. As I quoted at the beginning of this chapter, the letter included the lines:

We weaken our greatness when we confuse our patriotism with tribal rivalries that have sown resentment and hatred and violence in all the corners of the globe. We weaken it when we hide behind walls, rather than tear them down; when we doubt the power of our ideals, rather than trust them to be the great force for change they have always been.[19]

People will never all agree. Life would be boring if they did. Discourse, dialogue, and debate do not mean irredeemable differences. "Tribal rivalries" enrich no one. McCain is right—they sow resentment. Hiding from the enriching exchange of ideas makes us poorer. When we overcome the walls that separate us, we grow.

When I was a child and visiting my aunt and uncle in New Jersey, I jumped off the ledge. The quarry water rushed up to meet me as I fell. As I was in the air, I felt freedom. I felt a sense of being in the world that I would not have experienced if I had stayed behind the imaginary line at the end of the diving board. We can have that freedom in Christ. We can step out in faith, leave the safety of the ledge, and trust that God goes with us.

Notes

1. John McCain, "John McCain's Final Letter to America," *The Atlantic*, August 27, 2018, https://www.theatlantic.com/ideas/archive/2018/08/john-mccains-final-letter-to-america/568669/.

2. Lewis Carroll, *Through the Looking-glass: And What Alice Found There* (London: Macmillan, 1871), 193. Cited in John Barton, *Reading the Old Testament: Method in Biblical Study* (London: Darton, Longman, and Todd, 1984), 104.

3. Niels Peter Lemche, *The Old Testament Between Theology and History: A Critical Survey* (Louisville, KY: Westminster John Knox Press, 2008), 18–20.

4. Claude Lévi-Strauss, *Anthropologie Structurale Deux* (Paris: Plon, 1958).

5. Gene M. Tucker, "Isaiah 1–39," in *New Interpreter's Bible*, ed. Leander Keck (Nashville: Abingdon, 2001), 27.

6. Carroll, *Through the Looking-glass: And What Alice Found There*, 193.

7. Paul Fiddes, *Past Event and Present Salvation, The Christian Idea of Atonement* (London: Darton, Longman and Todd, 1989), 50.

8. Pheme Perkins, "Mark," in *New Interpreter's Bible*, ed. Leander Keck (Nashville: Abingdon, 1995), 559.

9. M. Eugene Boring, "Matthew," in *New Interpreter's Bible*, ed. Leander Keck (Nashville: Abingdon, 1995), 420.

10. Leonardo Boff, *Jesus Christ Liberator: A Critical Christology for Our Time*, trans. Patrick Hughes (Maryknoll, NY: Orbis, 1980), 74.

11. Reza Aslan, *Zealot: The Life and Times of Jesus of Nazareth* (New York: Random House, 2013).

12. Flavius Josephus, *The Works of Josephus*, trans. William Whiston (Peabody, MA: Hedrickson, 1987), 748.

13. Fiddes, *Past Event and Present Salvation, The Christian Idea of Atonement*, 51.

14. Gustavo Gutiérrez, *Sharing the Word through the Liturgical Year*, trans. Colette Joly Dees (Maryknoll, NY: Orbis, 2000), 244.

15. Paul Tillich, *Dynamics of Faith* (New York: Harper, 1957), 116.

16. John Paul II, *Fides et Ratio: On the Relationship Between Faith and Reason* (Boston: Pauline, 2000).

17. Gerald O'Collins, *Revelation: Towards a Christian Interpretation of God's Self-Revelation in Jesus Christ* (Oxford: Oxford University Press, 2016), 106.

18 Fyodor Dostoevsky, *The Brothers Karamazov*, trans. Constance Garnett (New York: The Modern Library, 1996).

19. McCain, "John McCain's Final Letter to America."

||||||||||||||||||||||||||||||||||||||||||||||||||||

PART FOUR

Getting There

What does it look like to overcome polarization? Realistically, one person will not change the entire landscape. Possibly, one person can change her world. So, what does it look like to experience this change?

Throughout life there are doors and paths. We make choices about which door to open or path to follow. Each one has its own fragrance. Thinking about life in these terms reminds me of riding in the back of a pickup truck in Haiti. Various smells greet and assault the senses. The odors of putrid trash, rotting food, and stagnant water waft through the air. A breeze pushes away the rancid air, and the salt air of the Caribbean Sea refreshes the senses. Then, the faint smell of Haitian spices startles the senses as in the middle of the poverty, someone cooks a mouthwatering delicacy. Life is a mixture of fragrances.

This mixture represents the choices that surround people each day. Doors, paths, fragrances—these are all choices to a varying extent. When I am in Haiti, I might not be able to pick which scent hits my nose, but I can choose where to focus. Trash and rotting food are unpleasant, but I know that the sea breeze will soon bring a new scent. Even when a path feels fixed, there is some amount of human agency. When applying this idea of agency and choice to words, each person has the freedom to use any words in her vocabulary.

What are the words that guide us through each choice? The events of the Unite the Right rally in Charlottesville prompted the reflections of this book. My experience was as a bystander. I watched a skinhead yell some profanities at a counter protester, and the counter protester yell profanities back. They revealed both division and our shared humanity. Were they afraid? Of what? What does overcoming that fear look like? For each increasing division, what words matter? What can speak to the situation? How can we know we are making progress?

We can know that we are making progress when we react differently than we used to. When there is a change inside of us, we grow, unless it is a negative change like becoming fearful or depressed or angry. After the rally, a leftist teenager in our church grew frustrated because she wanted to "punch a Nazi," and I would not concede violence as an appropriate response. Am I right? Am I more mature in my faith? Probably not. She and I are at different places in our journey. Like her and every other person who has ever existed, I am a sinner in need of grace.

Constructive engagement with one another not only helps overcome barriers but also helps people feel that they are making progress in their faith journey. This action requires people to be strong in their faith. It requires listening, hearing, and contemplating what the other person thinks. We need empathy and openness and trust building. Seeking spiritual maturity helps overcome our connection to preconceived beliefs about God.

10
Showing Flexibility

"There is one body, and one Spirit, even as you also were called into one hope of your calling."

—Ephesians 4:4

"I shall cling to the rope God has thrown me in Jesus Christ, even if my numb hands can no longer feel it."[1]

—Sophie Scholl

In the introduction, we looked at Mary and the way blessing plays out in her story. Compared with the central role of Jesus in the Gospels, Mary is a minor character, but like many other minor characters, her significance exceeds the words devoted to her in the story. Imagining her story expands our empathy. Instead of a venerated saint, "Mary the Mother of Our Lord," what if she was a person who wondered, *If this is blessing, what is cursing?*

Mary shows how to be in the story but not take the focus off of Christ. Facebook and other social media do not foster focusing on Christ. Facebook does not help self-awareness or unity. The algorithm feeds division. What I "like" will appear again and again and insulate me from different perspectives. We live in a world of either/or responses. Getting there means recognizing the goal of being on God's side, not aligning theological precepts with a preconceived idea of the world. Each of us has our own theological precepts. These are our beliefs about God or the divine. When we align our beliefs about God with the way we conceive the world, we fail to allow God to transform our lives and seek to be our own agents of transformation.

Already and Not Yet

What does it mean to be strong? Long after Mary died and people established a Christian church, Ephesians 6:10 tells us: "Be strong in the Lord and in the strength of his power." What does that mean? Ephesians struggles between the already and the not-yet. When it was written, a few generations into the early church, Christ was already risen. Easter had happened. But Christ had not yet returned. Christians wondered what was next. They thought about what God was doing. If God incarnate had already risen, is there anything else? Or, if God was going to do something in the future, what would it be?

These early Christians were under all kinds of threats, so they looked to the future and the final destiny of humanity. In theology, this study of death, judgment, and the final end is called eschatology. Contemporary Christians face threats in a different way. As I noted, Facebook does not threaten Christian faith, but it presents a self-centered worldview as an alternative. Biblical faith seeks God first and puts the self second. Does God judge this present world? If so, is it even possible to make progress on a faith journey?

One of the later New Testament epistles, like Ephesians, can speak to the tension between this present world and the future. Ephesians moves between a future and a realized eschatology. A future eschatology means that God is going to do something. There will be an answer in the future. A realized eschatology means that God has already done something. Following the rhythm of Ephesians, we can see this tension. In Ephesians 1:20-21, Christ is seated above all powers. God's work is already happening. In 3:9-10, God reveals a mysterious plan for humanity through the church. God's work is already happening. There is a need for unity in the body (4:1-8) and structure in the church (4:11-12). "There is one body" (4:4). There is "one Lord, one faith, and baptism" (4:5). Different people have different gifts. Some are apostles and some are prophets (4:11). These examples present a realized eschatology.

In Ephesians 6:10-20, the figurative language of armor and weapons points to a future eschatology—what Raymond Brown describes as "the divine struggle with powers and the rulers of the present darkness continues."[2] For people who struggle with current political divisions, remaining strong in our faith allows us to continue engaging with what separates us. We can remain strong because there is something better in the future. God has already done something incredible in the resurrection, but there is still a bright future ahead of us. Along the way, passages like these verses in Ephesians encourage Christ followers to be strong and keep the faith.

Historical events can help provide context for contemporary issues. Consider a German Christian trying to grow in faith during the 1930s. Germany had a horrible economic and political meltdown in the aftermath of World War I. Hitler promised prosperity and renewed national pride. With the advantage of hindsight, it is hard to imagine his popularity even among some German Christians. We now know about the horrible atrocities his government committed. The Gestapo and S.S. repressed German citizens who tried to criticize the government.

One of the greatest threats to the Nazis' power was information. If more people knew what they were doing, then more people might resist. Beginning in June 1942, Sophie Scholl, her brother Hans, and some fellow students at the University of Munich formed a group called the White Rose. They chose words over guns. As a member of the group, Sophie Scholl has now become an iconic face of nonviolent resistance. The group produced and distributed leaflets on a hand-cranked mimeograph machine and sent them to randomly selected addresses. The leaflets criticized the Nazis for abolishing the democratic process and suppressing free will and religious beliefs. They accused the Nazis of betraying the German people. They told about the murder of Jewish people in Poland. At that time, they knew of 300,000 deaths. By 1943, they were in contact with other resistance groups.[3]

On February 18, 1943, Sophie Scholl and Hans Scholl were distributing leaflets at the university when the caretaker caught them. He was a committed Nazi and turned them over to the Gestapo. As a woman, Scholl might have been able to get a lesser sentence if she gave up the other members of the group. She stood firm. She denied her captors the satisfaction of her surrendering information. They held her body captive, but they could not capture her mind. Her courage demonstrates a small step in breaking down the barriers that divide people.

After four days of interrogation, on February 22, 1943, the notorious Judge Roland Freisler sentenced Scholl, her brother, and another member of the White Rose to death. The Nazi government executed them just a few hours later. Today, at Munich's Perlacher Forst Cemetery, just a few hundred meters from the Munich-Stadelheim Prison where she died, there are three large crosses to mark the final resting place of Sophie Scholl, Hans Scholl, and their friend Christoph Probst.

Showing Flexibility

Our world values flexibility, and to some extent, flexibility is important. We have to be able to go with the flow, to not be intransigent in all things. Being flexible is key to empathy, listening to others, and overcoming differences. We might have one idea of how we should do things; then, we find that someone else has another idea. If we lack flexibility, we will stand no chance of working together. Soon, God's kingdom will be further fractured. Too often, people have staked their lives (or church membership or political affiliation) on minor matters. We must not major on the minors.

Søren Kierkegaard writes: "Purity of heart is to will one thing."[4] We value flexibility, and it can unfortunately creep into our theology. We might say, "Let us surrender this point or that point and decide, for example, that we do not believe in sin anymore." Now, it is more like a spiritual faux pas. Or, the opposite is true. Some people may pick one thing and let their entire faith rest on it, such as believing one's views on abortion serves as a litmus test to their salvation. This kind of dogmatic faith is not a purity of heart either.

The idea of being strong in the Lord does not mean *be strong in what matters to you*. Spiritual strength is not an ode to inflexibility. We can work together. Our struggle is not against one another but against the cosmic powers of this present darkness. Sophie Scholl and her compatriots worked together for their common beliefs. Their present darkness was truly a shadowy force in Hitler and the Nazis.

We miss out on the work of God when we refuse to work with other Christ followers simply because our church mission statement differs from theirs. Dr. Seuss tells a prescient story that could be about Christians in the twenty-first century. The Zax are stubborn and refuse to go in any direction other than their intended path. Once a southbound Zax encountered a northbound Zax and neither would budge even an inch to let the other one get by. They cannot solve their issue and the world progresses without them as the years pass. A highway and city eventually grow around them. Still, they stand, face to face, angrily staring at each other.[5]

To some, a children's story can seem silly, but Dr. Seuss's simple story illustrates the need to be flexible and open with one another. Getting there means getting over ourselves. The interconnectedness of the world means that it is not about me, nor is it about you. All of us are connected. Stubbornly taking a stand and refusing to try and see the world from another person's perspective misses God's presence and activity in the world.

To experience the spiritual growth associated with getting there, we must prepare for the next step in our faith journey. Stand firm for what is right. Stand against what is transient. Stand for what is transcendent. After preparing, we participate in what God is already doing. God is present, active, and moving in this world. We do not originate God's work; we join with God.

Notes

1. Sophie Scholl and Hans Scholl, *At the Heart of the White Rose: Letters and Diaries of Hans and Sophie Scholl*, ed. Inge Jens, trans. J. Maxwell Brownjohn (Walden, NY: Plough, 2017), 283.

2. Raymond E. Brown, *An Introduction to the New Testament*, ed. David N. Freedman, The Anchor Bible Reference Library (New York: Doubleday, 1997), 624.

3. Frank McDonough, *Sophie Scholl: The Real Story of the Woman Who Defied Hitler* (Stroud, Gloucestershire: History Press, 2009).

4. Søren Kierkegaard, *Upbuilding Discourses in Various Spirits*, ed. Edna H. Hong and Howard V. Hong, vol. XV, Kierkegaard's Writings (Princeton: Princeton University Press, 1993).

5. Dr. Seuss, *The Sneetches and Other Stories* (New York: Penguin, 1953).

11
Preparation

"Illusions are more common than changes in fortune."
—Franz Kafka

"Then all those bridesmaids got up and trimmed their lamps."
—Matthew 25:7

"Religion that is pure and undefiled before God, the Father, is this: to care for orphans and widows in their distress, and to keep oneself unstained by the world."
—James 1:27

Divisions in the world are nothing new. Preparing for divisions and trying to find ways to overcome them are a challenge. Asking questions, being open, studying issues, developing friendships with people who think differently, empathizing—these are all ways to depoliticize a situation. Discussing a subject rather than arguing about it and deescalating a heated conversation foster constructive communication.[1]

Kumbaya moments deceive us. For a brief period, people get along and it seems as though political, racial, and socioeconomic differences diminish. Then, something happens to bring the differences back to the forefront. Barack Obama was the president of the

U.S. when a white police officer shot an unarmed black man named Michael Brown in Ferguson, Missouri. Before that shooting, some people might have felt that racism had declined. That shooting, and so many others like it, challenged the feeling of harmony.

President Obama was in the Oval Office, right? Yet, racism, divisiveness, and politicization had not gone anywhere. In some places, they were less obvious. In others, they were just as apparent as always. The chances are good that if one were to have asked Michael Brown before the shooting, he would have said that racism was just as loud as it had always been. The question becomes: if we want to be agents of reconciliation, what do we need to do to prepare?

Finding common ground can have a Kafkaesque quality to it. Kafkaesque is something that is reminiscent of the oppressive or frightening qualities of Franz Kafka's fictional world. Kafka was a Czech author who wrote in a unique style. The novel *The Castle* is brilliant, entertaining, and infuriating. The plot flows and the scene is vivid, yet the story never wraps up. The main character, who is known only as K, arrives in a village where he is supposed to work as a land surveyor. The rest of the story involves his struggling through a bureaucratic nightmare to reach mysterious authorities who govern from a castle. At every turn, he encounters setbacks. He thinks that he is supposed to do something, but he meets another hindrance or holdup. The worst part about *The Castle* is that Kafka died before finishing it. So, after 350 pages, the story stops. It does not end. It just stops.[2]

A Kafkaesque God

The Bible is full of Kafkaesque moments. Some are simply puzzling. Others have unexpected turns, but when we put ourselves in a story, it can feel like Kafka's frightening and topsy-turvy world. For example, the Parable of the Ten Bridesmaids (Matthew 24:1-13) is reminiscent of Kafka. The bridesmaids all expect to go to the banquet, but five run out of oil and the story punishes their unpreparedness. The story twists away from the expected outcome. They all meant to go to the wedding. They went out to wait. The parable shies away from explaining why five had oil and five did not—it just says what happened. Why would five bridesmaids be excluded from the banquet because they let their lamps go out? The story is strange. If God loved these bridesmaids, why exclude them from a divine banquet? The tone feels objective.

The parable of the ten bridesmaids takes readers straight to the obvious message: be prepared; think ahead; plan, take extra oil. For conversations with people who approach the world from a different vantage point, the lesson could be to get into the right mindset to have a potentially confrontational exchange. Keep your cool. Get your facts straight. Check your sources. This is a good takeaway but not quite the whole story.

With Jesus, stories have a deeper message. This parable is no exception. This story collapses the space between now and the time of Jesus, both the historical Jesus and the future eschaton. Time moves forward, yet we look in both directions, backward and forward. Someday the world will end, life will end. This is the eschaton, this final event. As we move closer to a promised end, we look back at the historical Jesus to better understand where the world is going.

Living Today for the Future

In the Gospel of Matthew, chapters 23–25 depict Jesus' judgment discourse. The speech begins with Jesus addressing the crowds and disciples. He looks at the present and explores the way God judges the present. Then, in chapters 24–25, he talks just to the disciples.

The theme of Jesus' judgment discourse is the future. How do we live today to be ready for the end of time? This is a big question. Søren Kierkegaard writes: "All who are expecting [something] do have one thing in common, that they are expecting something in the future, because expectancy and the future are inseparable ideas."[3] Jesus knows that people look ahead and expect something. In Matthew 24, we find the "little apocalypse," which is about the end of the age, foretelling persecution, and desolating sacrilege.

If there is no future, there is no past. Kierkegaard writes, "If there is neither future nor past, then [humanity] would be in bondage like an animal," with our heads "bowed to the earth and our souls captive to the service of the moment."[4] On the contrary, what we do today is connected with the past, present, and future. We can react to the past, live in the present, and think about the future.

We are not, as Kierkegaard writes, "captive to the service of the moment." We are capable of seeing God and responding. The Parable of the Ten Bridesmaids is unique to Matthew. Neither of the other two synoptic Gospels have it, nor does the lost source Q. It connects with the immediately preceding verses and maintains the theme and connection with the rest of the judgment discourse. Why do we care about the judgment discourse? Will it help us in work or school or at play? Yes, because the Parable of the Ten Bridesmaids is about life. It is about the kind of life we experience every day. It is about looking at our past and envisioning a bright future.

The Parable of the Ten Bridesmaids is not a discrete unit. It connects with the verses before and after, and it is part of a wider story, both the entire message of Matthew and the judgment discourse. Matthew 24:51 says, "He will cut him in pieces and put him with the hypocrites, where there will be weeping and gnashing of teeth." This is closer to Greek cultural notions of death than Hebrew theology. The thought continues in 25:1, "Then the kingdom of heaven will be like this."

Parables are supposed to be believable, but the details in this story are confusing. Why did the bridegroom arrive at midnight? Where is the bride? There is no mention of her. The story suggests that shops would be open at midnight, so the foolish bridesmaids could rush out to buy more oil. This assumption seems unlikely. Instead of a parable, this is an allegory, which is a significant distinction. A parable is a story that teaches a lesson. An allegory reveals hidden meaning. Jesus shares hidden meaning about the future and how we should live in the present.

Jesus shares hidden meaning through the image of the bridegroom who represents Jesus arriving at the eschaton. This imagery is clear from the rest of Matthew (e.g., 9:15; 22:1-3). The bridesmaids represent the church.[5] Petri Luomanen describes Matthew's community as a *corpus mixtum*, "a mixed body of both good and bad members."[6] Matthew does not think that the bridegroom's delay specifies a longer delay. Both the foolish bridesmaids and the wise ones were mistaken. The bridegroom (Jesus) arrives at the appropriate time. The bridegroom's arrival marks the end of time. Eugene Boring writes, "The futile attempt to buy oil after the arrival of the bridegroom, though historically unrealistic, shows the futility of trying to prepare when it is too late."[7] This is the same futility of learning how to do something (e.g., skydive, pilot an airplane, or drive a race car) after having already started.

We cannot prepare for encounters once they begin. We must do the work ahead of time.

Keep Awake

"Keep awake" is the final word in the parable, but they all slept, even the wise bridesmaids. In this *corpus mixtum* of today's church, both the good and the bad ones do sleep. No one is vigilant all of the time. Today, we fight to stay awake, to be present and mindful of God's work in our world and our lives. We are easily distracted by life in the twenty-first century. Consider mass shootings. Some suggest a link between vitriolic political discourse and increased violence.[8] Others argue that "more guns mean more deaths from crime and accidents."⁹ What about violent video games? What is the role of untreated mental illness? How does the erosion of nuclear families contribute? What is the answer? How can we respond, not just to the latest mass shooting, but to everything we encounter?

The answer is that this is God's world and we are in it. We have a calling, and we have the tools to react. We can bring peace to war and violence. We can speak words of love and reconciliation where we find division. Unlike K., the protagonist of *The Castle*, we can be prepared. K. had no chance to prepare and no idea how to prepare. Every step forward led to two steps back. His world kept changing, yet we can make progress. We can reduce gun violence. We can be agents of peace. We can live in a state of expectation and realization. We can be God's hands and feet. With hope and faith in Jesus' images of the Kingdom of God, we expect a beautiful future where we realize God is already at work in this world in spite of the warts of human sinfulness.

Accept God's Gift

What does the word "religion" mean? Does religion help people grow in their faith? Can it? Does it provide tools for engaging with people who approach the world from a different perspective? What does it mean to be religious?

James 1:29 says: "True religion is taking care of widows and orphans." Is it really so simple? Would taking care of widows and orphans disarm the polarized conflicts all around us? Many cradle-roll Christians have a warm affinity for religion. It is how we practice our faith. For millennials and "nones," the word "religion" might evoke something negative. They think of the hypocrisy of people who uphold institutions. Wilfred Cantwell Smith writes: "Jesus was not interested in Christianity, but in God and [humanity]. He could not have conceptualized 'Christianity.'"¹⁰ Christianity is a religion and it has a structure, but the structure carries the baggage of twenty centuries of religion.

In Greek, the word "religion" (*thrace-ki'-ah*) can be positive (e.g., Acts 26:5, "testifying about our religion") or negative (e.g., Colossians 2:18, describing a religion of "worshipping angels"). David E. Garland writes that "It is easy for religiosity to become so absorbed in the external routines of worship, the preservation of the purity of doctrine, and the veneration of beautiful worship buildings that it degenerates into a kind of 'churchianity' that disregards such things as justice, mercy, and faith."[11]

So, what do we worship? What is our religion? Are we part of the institution of church, or are we seekers of truth in Christ?

Religion is the way we relate to God and experience God's presence. One way to think about relating to God is through divine gifts. Each person receives many gifts. The mysterious subject in that passive-voice sentence is God. God gives each person many gifts, but people must be careful because accepting them carries a price. This is not a transaction, that is, we get the gifts only if we accept the price. The price is eternal but requires our complete commitment. People can accept the blessings or gifts God gives and ignore the price. We can ignore God's expectations.

The cost of the gifts of God is an unimaginable richness. The price is good (i.e., eternal) and transformative every day. We get a better life when we pay the price. When we become aware of the relationship between every blessing and God, we can take a posture of gratitude and begin to appreciate what we have. We can offer thanksgiving for all good gifts that surround us. It is not about religion—it is about God.

James connects giving thanks to encountering troubles. Troubles help us grow. Troubles are the spice of life. By comparing good times with troubled times, we can see when things are good and connect gratitude with trouble. It might feel counter-intuitive, but when we face difficulties, we can be grateful for easy times. In

1849, Fyoder Dostoevsky faced a mock execution. The tsarist police arrested him and 35 others, and they held them in bleak conditions for eight months. They sentenced him and 21 others death by firing squad, but the Tsar Nicholas I pardoned them in the final moments before his death. His troubles helped him see the world in a different way.[12] Although Dostoevsky's example is extreme, troubles can help us know when things are good.

Gratitude stemming from trouble is a step in faith. For James, faith is connected to action. Moving forward in faith by giving thanks to God for everything good requires actions. It is an intentional move. Martin Luther was frustrated with James because he was so focused on the centrality of faith. Luther points to John and the letters to the Romans, Galatians, and Ephesians as "showing you Christ and teaching you all that you need to know." He wrote, "Therefore, St. James' epistle is really an epistle of straw."[13]

But Luther misses the significance of James and the relevance James has for today's world. It is easy to do if we do not recognize James's intent. James was not making a defense for organized religion. The epistle begins with an assumption of faith in Christ and takes the question further. In other words, now that we believe, what are we going to do about it?

All About God

Just as the Kafkaesque allegory of the bridesmaids was about God, everything in James is about God and the way we relate to God. In James, "the course and quality of our lives matter to God."[14] When we recognize God's role in every gift or blessing, God asks something in return. James is like an advanced instruction manual for faith. Taking James and applying the lessons to our faith can embolden our preparedness for encountering the world. James helps us recognize God's role and find ways to fulfill God's purpose. Then, people who take the lesson to heart get to be "a kind of first fruits of [God's] creatures." Immediately after introducing God's expectation that people become "first fruits," James begins telling how.

How do we prepare for what lies ahead of us? How can we be ready to have a grateful heart? What does God expect from us? Are there widows whom we can help? Yes. Are there orphans who need us? Yes. Are there victims of poverty and exploitation who need us? Yes. Can we work together for a more just world? Yes. Are there bridesmaids who need to be reminded to prepare and buy extra oil? Yes.

In November 2017, I had a momentary break in a busy trip that helped me prepare my spirit for what was next. I was in Boston at a conference. The conversations and insights were rich as we discussed the nature of ministering in a polarized and politicized United States. The whole experience was a gift. On Thursday evening, after everything was over, I was waiting to meet a cousin who lives in Boston. I walked along the harbor, soaking in the history, when I happened upon the Armenian Heritage Park. In the middle, there was a prayer labyrinth like the one in the cathedral in Chartres, France. This one had a stone path, with grass along the sides. I started walking the labyrinth. The slow, methodical turns brought me into the presence of God and an awareness that everything I have comes from the author of life.

If we are open to God, we will hear the voice of the Lord saying, "Good. I am glad you are open. Now, get ready. I want to make you my first fruits."

Notes

1. Each one of the suggestions can be challenging. Stress, exhaustion, time constraints, poor diet, and little exercise—these and other factors can challenge constructive communication. Developing healthy habits makes it easier to maintain composure in a confrontational and emotional situation.

2. Franz Kafka, *The Castle*, trans. Mark Harman (New York: Schocken, 1998).

3. Søren Kierkegaard, *Eighteen Upbuilding Discourses*, ed. Edna H. Hong and Howard V. Hong, trans. Edna H. Hong and Howard V. Hong, vol. V, Kierkegaard's Writings (Princeton: Princeton University Press, 2015), 17.

4. Kierkegaard, *Discourses*, V, 17.

5. M. Eugene Boring, "Matthew," in *New Interpreter's Bible*, ed. Leander Keck (Nashville: Abingdon, 1995), 450.

6. Petri Luomanen, "Corpus Mixtum: An Appropriate Description of Matthew's Community?," *Journal of Biblical Literature* 117, no. 3 (1998).

7. Boring, "Matthew," 450.

8. Susan Philips, "Vitriol and Civility in U.S. Political Discourse: An Aftermath of the Giffords Attempted Assassination," *Anthropology Now* 3, no. 3 (2011).

9. Franklin E. Zimring, "Firearms, Violence and Public Policy," *Scientific American* 265, no. 5 (1991).

10. Wilfred Cantwell Smith, *The Meaning and End of Religion* (Minneapolis: Fortress, 1962), 106.

11. David E. Garland, "Severe Trials, Good Gifts, and Pure Religion: James 1," *Review & Expositor* 83, no. 3 (1986): 387–88.

12. Alexander Burry, "Execution, Trauma, and Recovery in Dostoevsky's "The Idiot,'" *The Slavic and East European Journal* 54, no. 2 (2010): 255ff.

13. Martin Luther, *Martin Luther's Basic Theological Writings* (Minneapolis: Fortress Press, 2012).

14. James L. Boyce, "A Mirror of Identity: Implanted Word and Pure Religion in James 1:17-27," *Word & World* 35, no. 3 (2015): 215.

12
Participating in God

"Because there is one loaf of bread, we, who are many, are one body."

—1 Corinthians 10:17a

"The idea of participation does full justice to the commitment of God to bodies, and to the real presence of the triune God in and through the sacraments...Paul surely intends this when he uses the phrase 'body of Christ' in an overlapping way of the resurrection body of Christ, the communion bread, and the church."[1]

—Paul Fiddes

"I was hungry and you gave me food to eat. I was thirsty, and you gave me something to drink. I was a stranger and you took me in. I was naked and you clothed me. I was sick, and you visited me. I was in prison, and you came to me."

—Matthew 25:35-36

What does it mean to be a Christian? One definition might include inviting people to walk down the aisle of a church and make a profession of faith. Then, we who are already in the church would baptize them. Is this what it takes to make a Christian? The Bible says something different. Romans 10:9 sets forth a pattern of profession followed by salvation. Yet, in Matthew 7:21, Jesus says, "Not everyone who says to me, 'Lord, Lord,' will enter the kingdom of heaven, but only the one who does the will of my Father in heaven." In James 1:29, true religion is taking care of widows and orphans. Micah 6:8 answers the question about what God requires of us, saying God requires us to "do justice, love kindness, and walk humbly with God." John 3:16 is another banner definition of salvation: "For God so loved the world that he gave his only Son, so that everyone who believes in him may not perish but may have eternal life."

Which is it? The Bible seems to be all over the place defining what it means to be God's follower. Who gets to decide which part of the Bible defines Christianity? Is my emphasis correct? Or, is yours? Can they both be? Why? Or, why not? Are some definitions mutually exclusive of other definitions? All of this further complicates understanding what it means to be a Christian.

Romans is a specific book, written to a specific group of people at a singular time and place. The same applies to Matthew, James, Micah, and the rest of Scripture. In each case, applying a particular passage as a universal maxim misrepresents the intent of the text. The Bible can point toward better understandings of God, but the Bible is not God. It points to God.

One clear depiction of final judgment comes from Matthew's Olivet Discourse. In Matthew 23–25, Jesus talks to the disciples on the Mount of Olives, telling them about the end times. In Matthew 25:31–46, Jesus presents a basis of salvation steeped in what people do. John A. T. Robinson describes Jesus' word as follows: "The vision of the Last Judgment with which St. Matthew concludes so magnificently the teaching ministry of Jesus stands out from the Gospel pages with a unique and snow-capped majesty."[2] This is a passage with some power, and it is a passage about salvation. It does not contradict other passages about salvation, but it presents a clear notion of participating with God in the world.

When exploring the reconciling of political differences, notions of participating with God in the world serve as blueprints. WWJD ("What would Jesus do?") was put on a line of bracelets and religious ware a few years ago. As sentimentalized or simplistic as a slogan might seem, asking the question centers one's focus on living as Jesus would have one live. The Olivet Discourse and the Sermon on the Mount (Matthew 5–7) show how God acts in the world.

Goats or Sheep

In Matthew 21, Jesus enters Jerusalem, and he spent four chapters irritating the Pharisees and everyone else in positions of power. In many ways, he careens toward his betrayal, mock trial, crucifixion, and resurrection. Following Jesus risks living in a way that goes against people in positions of power and authority. Matthew 25 is the climax of his final discourse. Eugene Boring writes that "It is not a parable, but an apocalyptic drama."[3] Parables begin with something familiar. Matthew 25 starts at the final judgment. For reconciling differences, the final judgment might create some common ground. In other words, we will all face death and eternity someday.

So, what is our awareness of our own actions and how God views them? When we do something, how do our actions reflect our beliefs? To go back to the question of being a Christian, does this mean that we believe in our salvation as a one-time event (making a profession of faith)? Or, is it something we continue to do? How does being a Christian connect with being agents of reconciliation? Does God want Christians to depoliticize the world? Should our lives reflect our faith?

What is our awareness of our salvation and how each one of our actions reflects it? When we read the news and hear about increasing divisions, we become aware of the need for reconciliation. Jean-Paul Sartre writes, "All consciousness is consciousness of something."[4] We are conscious of being a follower of Christ. We are conscious of the need to overcome differences.

According to the story of the sheep and goats in Matthew 25:31-46, following Christ includes concrete behavior. Here, in a high Christology, Jesus sits on the throne:

> "When the Son of Man comes in his glory, and all the angels with him, then he will sit on the throne of his glory. All the nations will be gathered before him, and he will separate people one from another as a shepherd separates the sheep from the goats, and he will put the sheep at his right hand and the goats at the left. Then the king will say to those at his right hand, 'Come, you that are blessed by my Father, inherit the kingdom prepared for you from the foundation of the world; for I was hungry and you gave me food, I was thirsty and you gave me something to drink, I was a stranger and you welcomed me, I was naked and you gave me clothing, I was sick and you took care of me, I was in prison and you visited me.' Then the righteous will answer him, 'Lord, when was it that we saw you hungry and gave you food, or thirsty and gave you something to drink? And when was it that we saw you a stranger and welcomed you, or naked and gave you clothing? And when was it that we saw you sick or in prison and visited you?' And the king will answer them, 'Truly I tell you, just as you did it to one of the least of these who are members of my family, you did it to me.' Then he will say to those at his left hand, 'You that are accursed, depart from me into the eternal fire prepared for the devil and his angels; for I was hungry and you gave me no food, I was thirsty and you gave me nothing to drink, I was a stranger and

you did not welcome me, naked and you did not give me clothing, sick and in prison and you did not visit me.' Then they also will answer, 'Lord, when was it that we saw you hungry or thirsty or a stranger or naked or sick or in prison, and did not take care of you?' Then he will answer them, 'Truly I tell you, just as you did not do it to one of the least of these, you did not do it to me.' And these will go away into eternal punishment, but the righteous into eternal life."—Matthew 25:31-46 (NRSV)

This passage is unique to Matthew, and it begs some questions. Who are all the nations? Does he mean all nations on earth? Does he mean people groups who have never heard of him? Does he mean people from other religions? Or, does he have a smaller group in mind?

Maybe he has Jewish and Gentile Christ followers in mind. The grammatical shift from the neuter *eth'-nay* (nation or people) to the masculine pronoun *ow-tos'* (them) in 25:32 shows that the judgment is about individual human beings, not nations as political entities. Jesus is talking about our judgment, not the judgment of a nation. This is personal.

When we think about Jesus distinguishing between the sheep and goat, who are the "least of these"? Did Jesus mean just his poorer followers? Or, did he mean something more universal? Boring writes: "The fundamental thrust of this scene is that when people respond to human need, or fail to respond, they are in fact responding, or failing to respond, to Christ."[5] When we feed the hungry, welcome strangers, visit the prisons, take care of the sick, and give clothes to those who do not have any, we do it for Jesus. We do it for Jesus, because we do it in his name. When we do not do these things, when we ignore people's needs, we ignore Jesus, and we may not know we are doing this, because he might be disguised.

Some individual Christians and churches help widows and orphans. They help people locally in their communities because the Bible tells them to. Beyond their communities, they do not engage very much. They would see themselves as sheep, helping Jesus in disguise. However, what about Luke's parable of the Good Samaritan (Luke 10:25-37)? The story answers the question, "Who is my neighbor?" and extends the definition of neighbor beyond national borders. Immigrants, people in other countries, and those who do not fit into neatly defined categories are neighbors and widows and orphans in need of God's care. God's definition of neighbor is generally wider than ours.

Righteousness

The righteous stand out not only by their actions but also by their attitudes. Matthew de-emphasizes the self-confidence of the righteous. They do not know they have been righteous. They have not done anything to earn their salvation. The inner working of God, their consciousness of themselves and God's expectations, led them to take certain actions in their lives (e.g., feeding the hungry, clothing the naked). These actions lead them to take action, and these actions lead to their divine acquittal. They take the action not to earn their salvation but because of their consciousness of Christ.[6]

A self-righteous person who understands that she is on the right side of history is almost intolerable. This imaginary person has trouble abiding by other perspectives, even those that are ill-informed or ignorant. The self-righteousness can be overt ("You don't get it"). Or, it can be muted ("Let me explain this"). In either case, the person imagines a future when her position will be widely understood as the correct one. People with this kind of self-righteousness see themselves as confidently as someone who stood against George Wallace during his "Segregation now and forever" speech. However, in real time, making a courageous stand against

140

tradition can be challenging. Rather than being self-righteous, the person can be gracious and seek to be Christ-like.

Where are we in Jesus' story about the sheep and goats? We constantly make plans for the future. If we are going on a trip, we must make reservations or check out the car. When we have a consciousness of God's expectations of us and presence in the world, we can look ahead to the future. We can make divine plans. We can ask ourselves: how does my life reflect the salvation described in this passage?

We must resist putting things in the Bible that are not there. In Matthew 25:31-46, Jesus does not talk about when this will happen. He does not talk about salvation in terms of making a profession of faith or baptism. These are responses to something God is already doing. What Jesus does talk about is our actions here in this life. He uses a common reference for his listeners: sheep and goats. If you have done something good, you are a sheep—well done, good and faithful servant. If we have ever ignored a need, we are goats—we are accursed and have an eternal pit of fire waiting for us. The truth is, there are times when we are all sheep, and there are times when we are all goats. The decision we have to make each day is whether we want to have a sheep day or a goat day. Both the sheep and the goats were surprised on judgment day.

Feeding the Hungry

Earlier I told Larry DiPaul's story. One day, I was with Larry in Camden, New Jersey, when a homeless person walked up to us and asked for money to buy food. Larry said, "I can give you a peanut butter sandwich."

The man accepted the offer, and Larry told him to wait while he went inside to make a sandwich. I said, "Larry, how do you know the guy isn't taking advantage of you?"

He said, "How do I know he's not? I can't afford to risk denying Christ a sandwich when he's hungry."

In similar ways, we can be conscious of Christ every day. We can be aware of God's expectations on our lives. Some days, we will be goats. Other days, we will be sheep. Maybe we will do some good someday for someone who does not deserve it. Maybe we will do something for Christ in disguise. We can stop worrying about who is a Christian and who is not and leave that to God. God loves everyone.

Notes

1. Paul Fiddes, *Participating in God, A Pastoral Docrine of the Trinity* (Louisville, KY: Westminster John Knox, 2000), 282.

2. John A. T. Robinson, "The 'Parable' of the Sheep and the Goats," *New Testament Studies* 2, no. 4 (2009): 225.

3. M. Eugene Boring, "Matthew," in *New Interpreter's Bible*, ed. Leander Keck (Nashville: Abingdon, 1995), 455.

4. Jean-Paul Sartre, *Being and Nothingness: An Essay on Phenomenological Ontology*, trans. Hazel E. Barnes (London: Routledge Classics, 2008), Section V. The Ontological Proof.

5. Boring, "Matthew," 456.

6. Sigurd Grindheim, "Ignorance Is Bliss: Attitudinal Aspects of the Judgment According to Works in Matthew 25:31-46," *Novum Testamentum* 50, no. 4 (2008).

13
God Sees Me

"Where can I go from your spirit?
Or where can I flee from your presence?
If I ascend to heaven, you are there;
if I make my bed in Sheol, you are there.
If I take the wings of the morning
and settle at the farthest limits of the sea,
even there your hand shall lead me,
and your right hand shall hold me fast.
If I say, 'Surely the darkness shall cover me,
and the light around me become night,'
even the darkness is not dark to you;
the night is as bright as the day,
for darkness is as light to you."

—Psalm 139:7-12

God knows us. God searches the depths of who we are. God is with us when we sit down and when we get up. God looks at our paths before we travel them. Even before we say something, God knows it. This is especially troubling because we do not always say words of love. We do not always say constructive things. How can this be if God knows what we are going to say before we say it? The psalmist does not say God intervenes, just that God sees.

So, what is the point?

James says the tongue is a fire. It stains our bodies. No one can tame it. The tongue is "a restless evil, full of deadly poison. With it we bless God and curse those who are made in God's image. The same mouth blesses and curses" (James 3:8-9). This is the mouth God sees. This is the life God watches. James says that we should not act like this (James 3:6-10). Our words and our actions should bring God glory, yet we still say things that tear one another down. Why?

In James 3:11, we have an example of the kind of words God hears us utter: "The same spring cannot produce both fresh and brackish water." Once, on a trip to the Everglades, I brought a backpacking water purifier. I knew about the need for fresh drinking water and did not want to have to carry a week's worth of water in my canoe. When the ranger asked how much water we were taking with us, I said, "Even if we run out, it's no problem because I have a water purifier."

The ranger pointed out that unless the water purifier was capable of desalinating water, it would be useless because much of the outer Everglades are brackish. The same body of water cannot hold both brackish and fresh water. This life that God watches cannot be both holy and unholy at the same time. We can have the sense of God's absence, but Psalm 139 reminds us that we are not alone. When we speak, God listens—even if we are not speaking to God. When we speak or listen, God is present. If we are impatient or apathetic, we deny our role as God's hands and feet in the world.

What We Face

God knows what we face. In Psalm 139, God goes beyond the cosmic creator who puts the moon and stars in place. We discover an inescapably personal God. The psalmist prays, "You formed my inward parts; you knit me together in my mother's womb" (Psalm 139:13). Almost every part of the passage includes a "you" or "your" and an "I" or "me" or "my." James Mays writes, "Psalm 139 is the most personal expression in Scripture of the Old Testament's radical monotheism. It is a doctrinal classic because it portrays human existence in all its dimensions in terms of God's knowledge, presence, and power."[1] To know that God knows us means that we matter. It means we can relate to God, not as a divine-other, but as a loving creator. The familiarity of a monotheism with a loving God makes it feel less radical, but a God who loves us and has experienced every aspect of our humanity can identify with us.

Where was this personal, interventionist God during the Holocaust? What about 9/11? The Rwandan genocide? There are so many times when an interventionist God would be especially helpful, but bad things continue to happen. God did not cause the Holocaust, 9/11, the Rwandan genocide, or any other bad event in history. Nor do those horrific events undermine God's power or love for humanity. Immanuel Kant writes, "Examples are the leading-strings of the power of judgment."[2] In other words, examples help us understand and support our point or belief.

We can find examples for God's intervention. If we look, we can find examples to support the personal intimacy with God that we find in Psalm 139. We could say:

- God watched over me on my journey (139:3).
- God guided my words to be beautiful and useful (139:4).
- God heard my prayers and healed my friend (139:5).

In the Everglades, the ranger prevented my dehydration by pointing out the need to carry enough fresh water. Was the ranger God's agent, or just a person doing her job? Psalm 139 does not say God intervenes. The problem is not finding examples to support the notion of God intervening, but the counterexamples that undermine personal intimacy with God. The psalmist says, "You search out my path." Yet, stories like Annelies Marie Frank are commonplace. People today know her as Anne Frank, the teenager who kept a diary while hiding from the Nazis. She did nothing to deserve their hatred. When the Nazis discovered her and her family hiding, they arrested them and she died in a concentration camp. The psalmist says, "You search out my path." But we may ask, Why mine? Why not save Anne Frank? Why not stop the Rwandan genocide or the California wildfires?

The journey continues, and the answer is not simple or cut-and-paste. Figuring out how to walk with Christ in one situation will not necessarily apply to another one. Overcoming a political difference with one person will not universally apply. Each situation is unique. Each one requires prayer, fasting, study, and dedication to find paths forward. I say "paths" because there is not always one unique solution. God works in different ways in different times and places.

God's Presence

I do feel God's presence. I feel that I can testify to God's watching my path. God watched me in the Everglades while traveling for fun, and God watched me on a short-term mission trip in Sri Lanka where I learned more from the people I came to help. My testimony pales against the counterexample of Anne Frank. Friedrich Schleiermacher describes God-consciousness as the "feeling of absolute dependence" on God.[3] What does it mean to be "absolutely dependent" on God? Was Anne Frank not dependent enough? Certainly not!

There is no correlation between God's intervention, engagement, and good things happening in life. People do not get cancer because God is punishing them. Babies do not die because God causes their deaths. Being dependent on God is part of spiritual engagement. The psalmist reminds us that we do not walk alone: "You hem me in" (Psalm 139:5). God is with us.

Even though the Bible says God is with us, we live in fear. Robert Gordon writes, "Fear motivates avoidance of vulnerability, so that if one's fears *prove true*, one will have salvaged what is importantly at stake."[4] Fear is universal. Even those who know that God is with them can be afraid. Even those who appear to be fearless will have some Achilles heel, some aspect of their lives that they would desperately like to avoid. It is this fear that leads us to shrink back from the devotional courage of this psalm. It allows us to point to deaths, like Anne Frank's, and say, "See? God is not really searching out my path. I should be afraid."

God does not call us to be afraid. God calls us to be courageous. There is much we do not know about Anne Frank's life, but each day is a gift. It is a day to be alive and to exhibit courage, knowing that God stands with us. We stand before God, not a distant, unengaged God who set the world in motion and then walked away. We stand before the Lord, who searches us and knows us. God sees us in all things, in all situations, in the good, in the bad, during celebrations and abysmal failures—and I can look at my life and see both.

Courage Is Complicated

Courage is complicated. It takes courage to try and to overcome polarization. It takes courage to speak with someone who approaches the geopolitical landscape from a different point of view. When we engage, we are often closer to common ground than we realize. Someone who is distinguished in one area can be an utter coward in another.[5] We might not be able to undo some of the bad. We cannot bring back someone who is gone. But we can join God in celebrating everything good. We can follow God and speak good, encouragement, and love.

As I wrote this book, I thought about some of my own highs and lows. One of the great high points of my life is the birth of my children. In Psalm 139, I see that God knew my children before they were born. God was present in the miracle of their formation. God was there, celebrating with me, in the moment of their birth.

Then, the universality of this psalm sinks in. Gradually, as I let it, the universality of the human experience sinks in.

God was present and celebrating at the birth of every child ever born.

God knows us. God searches the depths of who we are. God is with us when we sit down and when we get up. God looks at our paths before we travel them. Even before we say something, God knows it. God does not force our actions or choose our words. God knows us and wants a relationship with us. Do we want to be agents of change? God gives us the freedom to choose.

Where Now?

One big news event seems to follow another. Each story provides an opportunity to engage with others and listen to different perspectives. When the wildfires were ravaging California, I could ask how to discuss the natural disaster and depoliticize the conversation. It was a continuous exercise. When the hurricanes, and then later earthquakes, hit Puerto Rico, I could prayerfully consider how to get involved. When President Trump's impeachment inquiry started and then the U.S. House of Representatives passed the Articles of impeachment, I could pause and listen to hear God's voice. Huge events need the kind of pastoral response that any follower of Christ can provide. We can ask what God is saying and how we can reflect the love of Christ and God's grace.

Looking back to 2017, being more active against the Unite the Right rally would not necessarily serve a larger cause of bringing God's peace into the world. To some extent we are all bystanders, and the challenge remains to be engaged while maintaining connections. Do we jump in and alienate some people? Or, do we step back and risk losing our saltiness (Matthew 5:13)?

Opportunities to engage with people who see the world differently surround us each day. I was recently with some family members with a different worldview. I would classify them as fundamentalist, and they would classify me as liberal. Both summations misclassify the depth of the other person. In the past, we have experienced friction over political differences. When I was with them while writing this book, I tried a different approach. I listened more. I sought common ground, and I avoided the hot-button issues that would break down communication. Shared experience can be the basis for future conversations.

The most politically active sides in the Unite the Right rally suffered from a short view. A long view reflects thoughtfulness, relational depth, and a plan. The white supremacists must have known that their position was unpopular and untenable. If they are feeling dislocation, then it is incumbent upon them to find the words to express their frustrations. In a civil society, they cannot oppress another people group or take what they perceive to have lost by force. Those who stood against the hatred are brave. They should be celebrated, but their stand was also a short-run gain. Standing against hatred does not affect the hearts and minds of white supremacists.

Instead of settling into life as a bystander, each day presents an opportunity to grow. We can listen more, talk less, and be thankful for the chance to grow in Christ. Faith in God through Jesus Christ is not a destination; it is a journey.

Get off the bystander bench and get into the game. I went to the rally, and it inspired me to write this book. More significantly, the experience inspired me to try to be an agent of reconciliation and to work to overcome the politicization in America today. I hope you will join me.

Notes

1. See James L. Mays, *Psalms*, ed. James L. Mays, Patrick D. Miller, Jr., and Paul J. Actemeier, Interpretation (Louisville: John Knox, 1994), 425.

2. Immanuel Kant, *Critique of Pure Reason*, ed. Paul Guyer and Allen Wood, trans. Paul Guyer and Allen Wood, The Cambridge Edition of the Works of Immanuel Kant, (Cambridge: Cambridge University Press, 1998), 269.

3. Friedrich Schleiermacher, *The Christian Faith*, ed. Hugh Ross Mackintosh and J. S. Stewart, trans. D. M. Baillie et al. (New York: T & T Clark, 1999), 132.

4. Robert M. Gordon, "Fear," *The Philosophical Review* 89, no. 4 (1980): 560.

5. Nicolas Berdyaev, *Slavery and Freedom*, trans. R. M. French (New York: Charles Scribner's Sons, 1944), 160.

APPENDIX
Principles of Reconciliation

As readers consider how they may bring about reconciliation, they may want to ponder some essentials concepts that were discussed in this book. Too often, Christians produce:

■ **Hatred.** Disagreements become unacceptable. We cannot simply see something differently and discuss it. The other side becomes an enemy.

■ **Quarreling.** The most common fight in Christianity is among Christians.

■ **Disbelief.** "Nones" (those who describe their religion as "none") are one of the fastest-growing categories of religion, not because God has abandoned us, but because our relationship with God is not transformative but is transactional. It is as if people say, "I come to salvation not because I love God, but to avoid my perception of hell. Once I get that blessed assurance, I am ready to cruise through life." Yet, God did not make a deal. God offers a relationship.

Jesus recognizes this and wants us to model for the world what it looks like to get along with each other. The first step in this passage is the directive to go and seek out the person who sinned against you.

Jesus says, go to the person when the two of you are alone. Why? So that the person does not immediately feel defensive. Going alone sets the stage for mutual trust. It creates the possibility for real discourse. Arriving at different conclusions about some question does not constitute having someone sin against you. Just because you disagree with someone does not mean they sinned against you. Christianity is a fellowship of believers under Jesus Christ, and it includes different people from different traditions, races, nationalities, languages, and socioeconomic backgrounds. We can disagree and still break bread together.

Jesus invokes a tone of humility. Go to the other person humbly. Jesus says at the beginning of Matthew 18: "Whoever becomes humble like this child is the greatest in the kingdom of heaven." We go humbly and after having prayed. Use "I" statements, not "you" statements. Do it in person, and if possible, do it with food. Food is sacred. We ingest nutrients to stay alive. When we eat with another person, we are sharing an intimate act.

Never try reconciliation via written word. Words on a page risk misunderstanding. Vocal inflection and facial expressions are missing in emails, texts, or letters. Then, following the example of Matthew 18:16, if we do not reconcile, we go with another person. If we still cannot reconcile, we bring the matter to the church. Jesus says, "If the offender refuses to listen even to the church, let such a one be to you as a Gentile and a tax collector" (Matthew 18:17).

Does this mean cast the person out? No. Consider the way Jesus addressed Gentiles and tax collectors. Just as in Matthew, Jesus continually reaches out to a lengthy cast of characters who were considered unsavory. In Matthew 8, Jesus heals a centurion's servant. In Matthew 9, he calls a tax collector to be his disciple. When we think we are off the hook if we have tried to reconcile and been unsuccessful, we have missed Jesus' point. In Matthew 18:21, Peter asks about forgiveness, and Jesus says, "We do not just forgive seven times, but seven times seventy." In other words, we forgive over and over again. For Jesus, forgiveness and living in communion together comprise the Christian life. Reconciliation is hard work. It takes time and involves risk.